JUNIOR OFFICER

William P. Capps

ISBN 978-1-7378523-2-2

Publisher
CAPPSIZE CREATIONS
1449 Pavillon Drive
Hoover, Al. 35226
cappsize@aol.com
(205) 960-8504

This book is a fictionalized autobiography. It is mostly a truthful telling of the author's experience during an extended period in the United States Air Force. Many of the characters are discussed giving their real names and are so noted. Others are either a figment of my imagination, a failed memory or a composite of several different people who found a way to add to the narrative. Any resemblance to persons living or dead is entirely coincidental.

This book is dedicated to
Lieutenant Colonel Ralph Angus
and
Colonel Thomas S. Torresson, Jr.

both of whom strongly influenced
my time in the Air Force
but also
the rest of my life.

BEGINNING AN EXTENDED JOURNEY

TWAS THE NIGHT "AFTER" CHRISTMAS
AND IN THE MIDDLE OF THE NIGHT
TWO CREATURES WERE STIRRING
IT WAS QUITE A SIGHT.

Growing up I never learned what was etiquette for a situation like this. I had gotten no sleep at all on Christmas night and starting at 5:30 a.m. on the day after, I had gone through an exhausting induction physical for military service. It was not complete until late in the afternoon. I drove from Montgomery, Alabama to some little town in Louisiana where I had stopped for gas, ate a quick meal and checked into a cheap hotel. The year was 1962. We did not have the nice chain motels that we have today. Climbing to the fourth floor (no elevator) I had hit the bed and had fallen into a deep sleep. I was awaken at two in the morning by noises from across the room. Squeaking bed springs, moans and occasional unintelligible words as two naked bodies bounced on an adjoining bed.

There was an eerie yellow glow shining through a window from a sign on a building across the street. My traveling companion had apparently found a diversion from our trip.

Let me see if I can bring you up to date on how I got to this unusual point in my life. In May I had graduated from Auburn University with a BA-A Degree in Advertising Design.

OKAY I'M OUT OF SCHOOL NOW WHAT ?

When I was a child I had always thought that someday I would go into service. My dad and most of his brothers had served in the military during World War II and I wanted to do so also ….. but which branch? One of my dad's sister's husband was still in the Marines. He was career military and a full Colonel. He was one of the very few Marines who came all the way up thru the ranks. He had been captured at Corregidor during WWII and held captive for three years. We had visited them many times and I gave the Marines serious thought.
Yeah sure …... for about three minutes. I had read about Paris Island and didn't want any part of it. Anyway the Marine Corp ad said they were looking for a few good men so I planned to be one they wouldn't find. Another reason I needed to consider a branch of service was the draft. I was at

that age (22) for which I was eligible. Unlike many of my friends who married early and had kids, or had some other deferment such as medical , I had none. The Army was not going to be my first choice especially if I could not go in as an officer. I did have two years of Air Force ROTC at Auburn. It being a Land Grant College, two years of military ROTC were required. I had chosen the Air Force ROTC over the Army and Navy which were also offered, mainly because my Uncle John had worked with the Air Force for many years as a civilian at Bolling Field in Washington, D.C.. I hated going to drill. I also hated the book work which went along with it. All of the tests were multiple choice (and I have never done well with those). It seemed that the football players always got hold of the tests beforehand and those tests were graded on the curve. Therefore I think I always made “D”. Well, that is until my last quarter of ROTC.

I could not take Advanced ROTC because in Ad Design I had three hour labs every day. That last quarter of ROTC one of my labs actually overlapped the hour I was supposed to have drill. I’m not sure how the schedule was not caught when I registered for classes that quarter. I had someone answer roll call for me and I was never exposed. Also for the written tests I did not study and answered “All of the above” fifty percent of the time on the tests. That was the only quarter I made a "B".

I did look at the Coast Guard but heard that the physical for it was tough. I talked to their recruiter though. In talking to recruiters (which did include the one from the Army) , I asked each one if there was something I could do in service which would utilize what I had studied? Was there anything where my degree in Advertising Design could be of use? The only one who offered any encouragement was the Air Force Recruiter. He said "Well we could sure use you in Recruiting". Unfortunately he told me he did not think I could just go straight into a position like that. His suggestion was I go to OTS (Officer Training School). It was a program they had for college graduates where after only three months of training you were given a commission as a Second Lieutenant.

My friend (since the 7th grade) Bill Lollar had stayed in the Navy Advanced ROTC and graduated as an Ensign (same as a Second Lieutenant in the Air Force). I thought, WOW, he spent two more years getting his commission and I could do it in just three months. I was ready to sign on the dotted line. The recruiter gave me several brochures and some explaining the OTS Program. I thought at the time …. someone designed these why can't I do that. Even though I felt I was sold, I told him that I wanted to take the material and some from the other recruiters I had talked to and discuss it with my parents. He said that sounded like a good idea. The recruiter was a Master Sergeant. I did not know that at the time but would learn about all ranks in the not to distance future.

I went over all the stuff with my parents. It seemed to be a big decision but thinking back, my options were rather limited. I could sign up with the Air Force and providing I made it through Officer Training School I could spend four years as an officer. Or if I waited around to be drafted, I would be a Private in the Army. The draft was two years. The choice was really Army for two at the bottom or Air Force for four as an officer. I went back the next day and signed the papers. My induction was set for December 25, 1962 …… Merry Christmas from the U.S. Government. Santa's red and white suit was going to be replaced with red, white and blue by Uncle Sam. That old bag of toys would be swapped now for a duffle bag.

Scheduling me for Christmas day should have been my first clue that my time in the military would be anything but normal.

Early in November I had purchased my first new car. I had sold my used Edsel (which I had in college). Money from it and with the cash I had saved, I had enough for a down payment. It was my first real purchase “on my own”. I had convinced the car dealership that I was about to be gainfully employed by the United States Government. As an OTS student I would be paid the same as a Staff Sergeant so I could easily handle the monthly payments. At the time being paid as that rank sounded like a lot of money. I would find out later that most real Staff Sergeants, with allowances, time in grade, etc. made much more than the average Second Lieutenant. At any rate I purchased a 1962 Mercury Comet convertible. It was white with red interior and had a four in the floor transmission. Why I wanted a car that was not automatic I can only guess. At the time it probably sounded cool. Surprisingly I drove it until 1965 and never burned out a clutch.

I packed the car on Christmas morning. This did not take long as I was advised not to bring too much personal stuff with me. I had about 4 weeks worth of civilian clothes, plus a few odds and ends that I could have done without. I did pack a graduation gift my Aunt Louise and Uncle John had given me which would come in handy later . It was a Polaroid Camera. The Induction Center was at Gunter Air Force Base in Montgomery, Alabama. It was 94 miles away. Everyone going into the military for <u>any</u> branch of service was processed through the same Induction Center. The recruiter had told me that another guy from Birmingham, who was to be in my same OTS class was going to give me a call. He was a good bit older, married and had prior military service. He had enlisted in the Air Force right out of high school and apparently served as an airman for a three or four year hitch. He was discharged and while his wife worked he had gone to college on the GI Bill. Now that he had a college degree he was going back in to become an officer. He called and suggested that he ride with me to Montgomery and then we ride together to San Antonio, Texas where we would be going for OTS. He explained that we would both be given a mileage allowance and by splitting expenses I would come out ahead. It sounded pretty good to me. I would have some company plus this guy already knew the ropes. Hey. I may have had ROTC but this guy had been in the REAL AIR FORCE. We set it all up just before Christmas. He sounded okay on the phone. He said he had two small kids and hoped we could wait and leave after lunch so he could be there for his two boys to have Santa and all. I told him I was flexible and not all that anxious to report to Montgomery anyway. He suggested I pick him up at his house about 1:30 Christmas afternoon. Richard Peters looked older than he sounded on the phone. He was probably only in his mid-thirties but his hair was receding a bit in the front and it had a touch of gray. His wife was small and fairly attractive. She helped bring his stuff to the car. It was one big duffle bag, a large suitcase and two smaller bags. A lot more than I was taking. His wife walked back to the house and brought back a paper bag filled with food and snacks. I never did see the kids. Richard's very first comment to me was "I called your house to see if you were on the way and it sounded like your mom was crying. Is this the first time her little boy has been away from home"? Maybe it was the way he said it …. or possibly because he said it in front of his wife but right off the bat I decided I didn't like him. I tried to slough it off and said something awkward like No, I've been away in college for the last four years, she's probably upset because she doesn't know if I am ever coming back".

The drive to Montgomery was long. Richard did most of the talking. He told me about his Air Force experience and about how much money he had been making the last couple of years selling used cars. I wondered to myself, if he's doing so well why was he going back into the Air Force. He even made a strange comment about how much he hated Officers when he was in service and he was having a hard time believing he was on his way to become one.

The Induction Center was in an old barracks building … just like the ones you have seen in old war movies. From the looks of the place I believe most of the military budget went to Maxwell Air Force Base which is also in Montgomery. We reported to a grizzly old Army Sergeant, who did not appear to be too happy about working on Christmas Day. He scratched our names off a list and said, "Grab a bunk in the next room. You won't be processed until tomorrow morning. They will get your asses up at five a.m. to start your physicals. You can do whatever you want tonight but you're not allowed to leave this building". Great, there are around 50 guys crowded into a lone TV room about the size of a small bedroom. They were seated on a couch, two chairs, on the floor and standing. There is nothing to eat. The one vending machine has an "Out Of Order" sign and the Coke machine says "Will Not Make Change". Luckily I remember the sack in the car which Richard's wife said had food and snacks, The sergeant was not exactly pleased with me asking to go to my car but when Richard suggested we would share with him some turkey sandwiches he changed his mind. The night grew longer and longer. I had claimed a top bunk, not far from the entrance. Richard found a lower cot about six places down the wall. The bunks were just about all taken and it was a big long room. They were filled with mostly young men right out of high school. There was talking, coughing, sneezing, cussing, snoring and even a little crying. For whatever reason …. I could not sleep. I tossed and turned until around one a.m. At which time I got down from my perch and wandered out to where the sergeant was sitting with his feet propped on his desk. I asked him why he had to "stand guard" all night and he laughed and said, " You would be surprised what all I have to put up with baby-sitting inductees". He told me about breaking up fights, catching kids sneaking out to go home, even finding one guy with his girl friend in bed with him. We got to talking and before I knew it, my watch said 3 a.m. He said it was no sense in me going back to bed since they would be getting me up in a couple of hours. He pulled out a deck of cards and we played penny ante poker until five o'clock. I can't remember who won but it must have been me because when I was leaving he mumbled something about "You're luck is about to run out when your physical gets started". He did add "I'm really glad you joined me".

Every movie you have seen of guys about to be inducted into the military have been accurate. An old Jerry Lewis movie comes to mind. I expected to see him in the long line of men and boys standing in their underpants. From boxer shorts to jockey style you saw it all. We all wore nothing else. No shoes, no socks, no pants, no shirt … nothing but skivvies. It was an assembly line.. We walked from door to door and it seemed like they checked everything. In addition to the fickle finger of fate up the rear we also had to cough several times. Some of those coughing the night before must have been practicing. We thought we would be given shots (as that was the rumor) but we were told those would come later ….. and for those headed to the Air Force that would be at Lackland Air Force Base.

We wondered how much longer our physicals would last until a young man appeared with a silly grin on his face. He asked for all the Officer Training School guys to raise their hands. There were about 6 of us. He said, "Y'all have to have some other tests done. Go through that door and walk straight through to the door on the other side". Several heads appeared out of offices like they were expecting something. We quickly found out why they were watching because we were being marched straight through the secretarial pool. Today we could probably file a sexual harassment suit but back in '62 we had to grin and bare it. There were cat calls, and whistles and several smart remarks. I remember a lot of laughing from a crowd peeking in the door from which we had entered. Richard had been bringing up the rear and that's exactly what he showed them as we went out the far door. He calmly pulled down his shorts and gave them a moon shot as he walked out. He said "Kiss it good-bye girls ...we're going to fight for your country". We gave him a round of applause and I believe I heard among the laughing all the girls did too.

At the time I was rather embarrassed but after it was over I guess it was pretty funny. Richard said that they probably try to pull that on all the college fellows heading to OTS. When we got out the other door, we found we were finished and all we had to do was get dressed and take our oath into service. It was given in a large room to all the inductees at the same time. We got separated from the other four guys going to OTS and would not meet them again until we got to Lackland Air Force Base in San Antonio.

I FORGET THE ALAMO BUT REMEMBER THE TRIP TO SAN ANTONIO

I spent what little time I knew Richard Peters, trying to decide if I liked him or not. What he pulled in front of the secretaries was pretty cool and now his past military experience was coming in handy. After taking our oath all the enlisted guys were told exactly what to do. The OTS guys were just handed a set of orders and expected to know how to take care of ourselves. Richard read them and said “Boy this is great, we don’t have to show up at Lackland Air Force Base in San Antonio for four days. Plus we get per diem and mileage”.I was not exactly sure what all that meant except I knew if we both got paid the same, and split expenses, I would come out ahead.

We got a late start leaving Montgomery and hit what little 5:00 traffic they had back then. We decided to just head west until we felt like stopping. It was probably around 9:30 or so when we hit some little town in Louisiana. We found a restaurant open and a service station (remember those? They still had them back in the sixties), but most everything else was closed. The restaurant was still serving and they said they would be open long enough for us to eat but if we needed gas we had better fill up because the station would close at 10:00 and there was not another for miles. Richard slid into a booth and told me to go get gas and he would look over the menu.

When I got back to the restaurant Richard was talking to an older blond waitress. She looked like she was left over from the fifties (it had not been that long ago) but she did have a bee-hive hairdo. Her name, honest to God was Alice, although that name and restaurants were not synonymous at the time. Richard was engaged in a hot conversation about where we were headed and she seemed impressed. He told me that we ought to stay at the hotel next door. Alice had told him that it was real clean and not expensive at all and we would not find another place with vacancies for a couple more hours. I had not seen anything which looked like a hotel when I went for gas but figured maybe I just missed it. The dinner was not bad, a little greasy but we were hungry enough to not care.

How I missed the hotel was easy to see. It had a small sign on the side of the building which was not illuminated at night. The lobby was a desk and two chairs with a dark hallway behind and a set of stairs to the left. The image is still vivid to this day. The entrance way and total lobby may have been 15 square feet. The man behind the desk was awake and appeared to be reading the stock section of the paper ….. which did not quite fit the image of his surroundings. He indeed had a vacancy as it was advertised, CHEAP. We climbed to the fourth floor. The building probably was not wide enough for an elevator. The room can only be described as vintage Civil War, high ceiling, big enough for two double beds, an antique dresser and one of those huge French armoire. There was no closet but the door I thought might be one was locked and maybe it was a connecting door to the next room. Two windows were on one side of the room. I sort of expected the bath to be down the hall but we actually had one. It too was large. It had a tub up on legs but no shower.

With no sleep Christmas night and a long day, I was so tired that I hit the pillow and was gone. Just before I climbed into bed, Richard said he was going downstairs to find a paper to read in the john. He also added that he might relax in the tub for awhile. As I told you at the beginning of this book I was awakened around two a.m.. There moving up and down in that weird yellow light was that same rear end which had been proudly displayed at the Induction Center. I could not believe my eyes. Richard was banging the hell out of the waitress from the restaurant. At the time I didn't know it was her but that was my guess especially when I saw an object which looked like a beehive on the dresser. That night had to go down as one of life's little moments. What do you do? Pretend to sleep through the whole thing? Hell, I had led a pretty sheltered life up until then. I did not know what to do. Should I say something funny like "Gee Richard I didn't know you were getting up early". Or "I've heard of room service but this hotel goes above and beyond". I did not have long to decide. Richard let out a loud yessss! And rolled over. I saw a naked figure get out of bed and head in my direction. My first inclination was to pull the cover over my head. I tried to play it cool and act like this was normal. She said "Are you awake"? I mumbled something to the effect of "I imagine the whole damn hotel is". She laughed and said "Well are you ready to have a good time"? I thanked her but said that I thought I would pass on the offer.

Alice spent a long time in the bathroom but she left before daylight. Richard never heard her leave. I got up around 7:15, shaved and took a bath. I did not expect Mr. Peters to be up before noon but he was up when I came out of the bath. The first thing he said was "You told that woman you would pass on the offer. You do know she wasn't offering for Free". At the time I didn't hear anything about money and I never got up the nerve to ask Richard what he paid. I guess since I turned it down thinking it was free, I'm sure he thought that I believed whatever he paid was way too much. If he did think that, he would have been correct. The expression rode hard and put away wet would have been a compliment. With me being 22 at the time and her on the downhill side of 40, I thought she was over the hill. I was also thinking that Richard had a lot of nerve but at least I only had to put up with him until we got to San Antonio.

He said "Let's hit the road, we can make it to Houston. We can spend maybe three days and have some fun. I knew fun for him could mean anything. Actually stopping in Houston had been my idea. I had dated a girl named Caren Davis from Houston. She was a student at TCU but lived in Houston. It was only one date in the fall of '61 after a football game in Auburn. She was visiting a friend of a friend or something and I ended up on a blind date with her. I remembered that she was real cute and we hit it off. She had introduced herself as "Hi I'm Caren with a "C"...... so all evening everyone called her Caren with a "C". We had each written a couple of times so I wanted to try and call her if we went through Houston. Richard was agreeable because he wanted us to take our full travel allowance and arrive at Lackland on the fourth day.

Arriving in Houston in the late afternoon, we checked into a real motel. You must know that in 1962 motels were not as plentiful as today. Motel Six had not been invented yet, so nobody had left a light on for us. Where we stopped was not on par with a Holiday Inn but it was clean and in a good part of town. I called my friend to see if she could go out for dinner and maybe a movie. Richard had said not to worry about him as he would find something to do. I was only worried what or who I might find when I got back. Caren with a "C" sounded great on the phone and she actually acted like she was glad I called. I told her I knew it was short notice but I was on my way to defend our country so she could sleep safely at night. I thought that the least she could do was join me for dinner to wish me farewell. She said her dad was out of town and her mom did not like to be alone at night. But she would ask and call me back.

The phone in our room did not ring until a quarter of six. I figured it was for Richard. He had spent over an hour trying to call his wife. He finally got through to her but most of the time he had been flirting with the long distance operator, who by now not only knew who we were but where we were staying. Richard asked her where all the operators went for fun after they got off. The call, however was for me. Caren's mom had suggested that instead of us going out, that both Richard and I come over for dinner. A home cooked meal sounded great so we got directions, cleaned up a bit and headed to her house. Dinner turned out to be tuna salad ….. so much for the home cooked meal. I think after dinner we played Monopoly (the short version) but even so we stayed later than we intended. Nice evening but not the date I had hoped for.

Arriving back at the motel around 11:30 p.m. we found a note on the door that Richard had a phone message. I cannot recall if this was in the days before message lights were on phones or if that motel was too cheap to have had that feature, At any rate the note was from the telephone operator which Richard had spent so much time talking to earlier in the evening. She and a friend were going to be at a local night-spot where "the action was". She suggested that Richard and his friend (meaning me) meet them there. Richard immediately grabbed the phone book and started looking up the address of the place. I knew to suggest that I might not want to go would give him an excuse to borrow my car (something I had decided was not going to happen). We took off to "Earl's Hideaway. It was a typical loud Texas bar ….. one of many I would see in the next year an a half of being stationed in the Lone Star State.
Earl's was dark and crowded. Apparently it was the "In Spot". There were many singles but many more "hard tails" (Richard's definition of guys) than females. Outnumbered at least two to one. I

asked how we were supposed to know who these girls were? He said that the operator's name was Shirley or Sherry and he would find her. I have to admit for a married man , he knew how to work a room. It did not take him ten minutes of asking various women if they knew any telephone operators named Shirley or Sherry before he motioned me over to a booth on the side of the bar. I said asking "Women" because to me the ladies I saw in the bar looked a lot older. Well Shirley or Sherry turned out to be Cheryl but Richard thought that was close enough for government work. He just slid into the booth next to her and started his smooth line. Cheryl looked to be around 35, bleached blond, a little on the heavy side but she had a deep sexy voice.

I was still standing beside the booth when her companion slid over and asked if I wanted to sit down. I reluctantly sat next to her. We all introduced ourselves and began a conversation. The two women were drinking beer.; Richard ordered one and I ordered a Coke (I did not like the taste of beer back then and still don't). They kidded me about not being old enough to order beer and said "Don't worry about being carded in here". I let them know I was indeed old enough and just did not care for beer.

Cheryl was a divorcee with two kids. The woman sitting beside me looked older than Cheryl. She said she was not married, never had been, and never planned to be. It was dark but I could see well enough to tell she would not have any trouble accomplishing her goal. Her name was not one I remember or cared to at the time. Cheryl did most of the talking for the table. She knew many of the people in the bar and quite a few stopped by to speak to her. There were a couple of girls who talked to Cheryl that looked more my age and my speed but I was not in a position to break out on my own. The more beers which came to the table, the louder the conversation got. I have always thought I had a fair sense of humor and I got on a roll and soon had them all laughing. I'm sure everyone thought I was as drunk as my companions were becoming. It was getting close to 2 a.m. and Richard had long since "drank Cheryl pretty". In fact he was saying that she could be a movie star with that sexy voice of hers. She was all over him in the booth. Twice my date (or I guess I should say seatmate) said "We better throw water on those two". She had gotten such a laugh the first time that's why she repeated it. The second time it was loud enough for anyone within four or five tables to hear. It was obvious to me that Richard, or as his now intoxicated friend was calling him "BIG DICK PETERS", was going to make his second conquest of our trip. I was equally determined I would not be a party to his plans or of those of the woman sitting beside me….if she had any. I finally said that I was going to call it a night and head back. I asked Richard if he was riding with me, got up, told the ladies I enjoyed meeting them and prepared to leave. Richard jumped up and put his arm around my shoulders as we walked a few feet away from the table. He said "Look buddy, we both got sure things, what's your problem. I'm beginning to think you're queer or something". No one used the word Gay …. back in '62 it meant happy. I snapped back that he did not have to worry about that, but I just was not interested in messing around with either of his new found friends. There was a lot of "hang on a minute" and "don't leave yet" while he ran back and forth talking to Cheryl. Finally he said "just go on, I'm gonna ride with Cheryl and we'll see you later". I did not like the "We'll see you" part. I was not in the mood to a repeat of what I witnessed in Louisiana. I told him as forcefully as I could "Don't plan on bringing her to the motel".

As I headed back to the motel alone, I got rather turned around and in fact got completely lost. Like most guys who don't like to admit they might not have the best sense of direction I did not want to stop and ask for directions. Also remember it was after two in the morning. Fortunately I came upon an open truck stop. I decided I would go in and have a piece of pie and ask for the quickest way back to the motel. I was only there for maybe fifteen minutes and it turned out I was close to where we were staying. When I pulled up in front of our room there was a big red Buick parked in front of our door. I could not believe that somehow they beat me back. I opened the door slowly expecting to walk into a brothel. I was shocked to see Richard sitting on the bed. Cheryl was not in sight but her absence was explained when I heard a gagging noise coming from the bathroom. Richard was cussing because she had gotten sick all over him in the car and she was still throwing up in the bathroom. The "Best laid plans of mice and men" Richard's plans were all that got laid that night.

The sequel: I'm not sure if it was his conscience bothering him or the fact we were on our way to become Officers and Gentlemen, by act of Congress but Richard said "You know maybe it would be a good idea for you to follow me. I think I should drive her car and make sure she gets home". I followed as Richard drove her car to an apartment complex about 20 minutes from where we were. Going back we got lost again but managed to find it on our own. Not many people to ask directions of at 4 a.m.

I did not wake up until around 9:30. Richard was snoring away. I walked over to the office to see when we had to be out of the room. I think it was the owner of the motel who was working at the desk. Back then there were not that many chains and many of the motels and tourist courts were locally owned. Anyway, the guy and I started talking about where we were going and when he found out we were in the Air Force and where we were headed, he said "Don't worry about the time. We are not busy this time of year. Stay as long as you want". San Antonio was only about a four hour drive back in '62 so we did not leave until after 2 p.m. Richard had to wash his clothes from the night before.

We stopped several times and did not get to San Antonio until well after 6 p.m. Richard was determined we were not going to check in until our fourth travel day so he said we ought to find a place to stay in town. The Crocket Hotel is a landmark. It was actually built in 1909 on the site where the siege of the Alamo took place and still stands (many times remodeled) today. I think it was actually called the Davy Crocket back then because that's who it was named after. Anyway we decided to stay there. Since we had saved so much money I suggested separate rooms. The next day we explored San Antonio (Richard had been there before) and spent another night at the Davey Crocket. I would return to this hotel after graduation from OTS but for an entirely different reason.

LACKLAND AIR FORCE BASE

It lacked a lot of things but "Land" was not one of them. We drove onto the base early in the day.

Signing in was the easy part. I was told to take my car to a large parking lot about a mile away from where we were getting registered for Officer Training School. I made my first of many mistakes when I asked how I was to get back. "You run back !, and bring anything you think you will need from your car because don't plan on seeing it again for around a month". Richard had already gotten all of his stuff and he was on his way to a different set of barracks from where I had been assigned. He did not say good-bye, thanks for the ride, good luck, kiss my butt or anything. He filled out his papers, got his barracks number and was gone. I thought I would probably see him someplace later but never saw him again. Luckily, I had suggested we split expenses as we went, so at least he did not disappear owing me any money.

The barracks were nice, freshly painted and were not at all what I expected. I thought they would be open bay type with bunk beds lined on both sides like we had at the induction center. It turned out we had individual rooms, similar to a college dorm. Most of them had only two bunks and two desks. Clothes closets and drawers were built in the wall. The room to which I was assigned was the exception for our barracks. Naturally I would get the exception. It had one single bunk on one side of the room and a stacked set of bunk beds on the other. It was the only room in the building where three men would be assigned instead of two.

Since I had jogged back from the parking lot, I was the last to arrive so I got assigned to the room with two other guys. The only bed left was that top bunk. Although this was not obvious at the time, it turned out to be a blessing in disguise. It was easier to make up the top bunk than the lower one, plus inspecting officers were more lenient checking the top.

OTS was divided into two phases: the first one and a half months you were considered underclassmen and basically your life was made miserable by those cadets in the last half of the ninety day period. For those who have been through fraternity and sorority initiations back when hazing was in vogue, the first half would be close to that experience, only much worse and for a longer period of time. We were told most of the really bad stuff was copied after what the Air Force Academy cadets put up with during their freshman year. We did have some Air Force Officers assigned to the OTS program who had graduated from the Academy. We would find out after we got into the real Air Force that Academy graduates were easy to spot. There was always an air of superiority. They felt they were better than the Air Force Officers who received their commissions through ROTC, Officer Candidate School or especially through OTS. They called us "Ninety Day Blunders" Thinking back they probably had good reason to feel superior but we still hated them for it. On most bases they were outnumbered anyway. Even so their description ended up as a good subtitle for this book.

I mentioned Officer Candidate School (OCS). This was the method of becoming an officer through the enlisted ranks. Most of those obtaining their commissions though this source were career people who had shown leadership ability as non commissioned officers. Some of them had completed college but it was not a requirement as it was for OTS. The majority of officers who I knew who had received their commission through OCS were dedicated, hard working, and seemed to get along well with everyone they worked with. I'm not sure who decided to eliminate this method of obtaining a commission as it appeared to be a good source for new officers. I think someone up in Washington came to the conclusion that all Air Force Officers should have a college degree. In my opinion, some of the practical experience the OCS candidates had was worth more than that piece of paper called a degree. At any rate the last OCS class was going on at Lackland at the same time as my OTS class.

The OCS cadets and the OTS cadets were kept separate but would occasionally see each other when marching or jogging from classes or the drill field. For no particular reason there was a sense of animosity between the two groups. We were made to think they had it rougher than we did but I doubt it was much different. At the time I did not think we could have had it any worse. To say I hated my first week would be the greatest understatement of my life. I could not believe I had gotten myself in such a mess. There was an "OUT" and we were told of it daily. It was called SIE which was the abbreviation for "SELF INITIATED ELIMINATION". If things got too tough you could decide, on your own to just quit and go home. There was a slight catch. If you were my age and my draft status, chances were good, you would come back to Lackland as an Airman Basic or head for the Army as a Buck Private. For all the OTS students who had prior service …. it was simple. They stuck around long enough to start catching hell, got fed up and told the Air Force where to stick it, and went home. I found out that was exactly what Richard had done. The first time one of the young upperclassmen told him what to do, he said adios and opted for the self initiated elimination. Being around him only that short period I was not at all surprised. Since I had too much to lose I decided I needed to roll with the punches and try to stick it out.

The decision to try to make the best of whatever the Air Force threw at me was not an easy one. I agonized over it for several days. There was a part of me that just wanted to admit I had made a serious mistake and the service was not my cup of tea …. at least as an officer. There had never been any question as to whether I would serve a military obligation. Growing up in the South where patriotism was an honorable thing, I had been raised with the notion that no matter what I planned to do for the rest of my life, I first would serve my country in the military. Today it sounds pretty hokey. I don't think the generations following mine have had the same inclinations. In a way that's too bad. It is almost the same kind of apathy which keeps people from exercising their right to vote. For better or worse I had joined the Air Force, and up to that point the "worse" was all I could see ahead of me.

I can clearly remember calling home and talking to my dad. He listened to my tale of woe and finally said if it got too rough to just come back home. I guess hearing I would not be disowned if I admitted defeat, was all I needed. When I hung up and went back to my room I was more determined than ever, to not only take all they could dish out, but to actually find some way to have some fun with the whole experience. I had survived "HELL WEEK" in a fraternity initiation and after it was over I had lots of stories to tell. Who knows , if I could survive this, maybe there would be enough stories for a book.

GETTING STARTED

Each new cadet was assigned an upperclassman who would be their mentor and give them advice. I was lucky enough to get one who acted as if he was truly interested in showing me the ropes. He was a graduate of Florida State and when he had heard that I went to Auburn, he had asked that I be the cadet assigned to him. The two schools had a good relationship and many of the students from each school would visit the other for parties and football weekends I had visited Florida State twice while at Auburn. This was back in the days BBB (before Bobby Bowden). Before he became head coach Florida State was the team you always scheduled for homecoming because you thought it would be an easy win. Starting in 1954, Auburn had played them 4 different times and won them all. The score in 1960 was Auburn 57, Florida state 21 and we had not played again until the '62 season so it was still fresh in both our minds. Florida state had tied Auburn 14-14. To Brad Gillum, my mentor's name a tie had been a win.

Brad and I hit it off immediately. He set out to make my OTS stay as painless as possible. He held the rank of Officer Trainee First Lieutenant. You could hold the rank all the way up to OT Colonel (and there would be only one of those in each half, upper and under). Brad explained that competition for the ranks was held during the first week of training and you held that rank until graduation. According to Brad, if you were serious about the Air Force as a career, you should try to get as high an OT rank as possible. I told him I was only planning on giving the Air Force the 4 years I had committed to when I enlisted. I wanted to get out and try to put my degree in Advertising Design to use. He said he felt the same way, saying ”You know I also have a weird degree”. I did not think of mine as weird but I wanted to hear his. He said “ I majored in Hotel and Restaurant Management”. I started to tell him about my experience running the fraternity kitchen but decided against it. He went on to give me some great advice. He suggested that I try to make OT First Lieutenant as he had done. This would at least put me above the masses (OT Second Lieutenant) and would give me a little rank and I would get to do some extra duties.

In showing me the ropes he wanted me to know a few tricks he had come up with. He said “If you want to maintain your sanity and get through all the bull shit, it is best to try and make a game out of it. Beat the system”. He said that he did and it drove his upperclassmen nuts. I had already decided I was clever enough to play the game and not let it get me down. The BS had already started before I met Brad. We were told the “RULES” by a little sawed off OT Captain by the name of Sammy Hitaller. It was pronounced Hi Taller, but as you can guess behind his back we soon called him "Little Hitler". OT Captain Hitaller was obviously impressed with his rank and thought we should be also. He explained how everything in our room had to be "grounded". Bed posts must touch base boards, lamps on desks must touch edges, chair when not in use must touch desk, etc., etc. Even the Venetian blinds had to be at a forty-five degree angle at all times. This was checked by sticking a pencil in them to be sure it stuck out parallel with the floor. Any violation would get you a demerit slip. Three demerits in one week meant you would be gigged in for the weekend. We had already been told at our signing in orientation that from noon on Friday until six p.m. on Sunday we would have Free time. This would include permission to leave the base. It was becoming obvious, since little Hitler was our inspecting officer that he was going to do all he could to freeze our Weekends. In addition to all the other Mickey Mouse, we were given instructions as to how our closets were to be set up. All shirts on hangers, facing the same way with sleeves folded inward. At this point we had not even received our issued uniforms but Captain Hitaller had a sample room already set up. Even the drawers had to be according to plan. Socks and underwear had to be rolled in a particular way and no matter what you had, it was to be in a certain place. Of course we got the expected instructions on how to make up our bunks, with hospital corners and blanket pulled tight enough for a quarter to bounce. As Brad had warned the BS was starting to pile up and would only be a small sample of what was ahead.

AIR FORCE ISSUE

Everyone should get to experience the Processing Center at Lackland Air Force Base at least once. I think it was called the White Monster or whatever color the building was at the time. We were all sent by bus over to a huge building about the size of several football fields. After filling out a few more forms, we had our hair cut .... or make that our heads shaved. We looked like a convention of Hare Krishna. I don't remember them being in existence back then (but I have been told they were). Filling out the forms was even an experience of its own. We were given instructions from an old Staff Sergeant who probably barked out the same commands so many times a day that they actually made sense to him. He said and I mean really fast "Fill out the top line, last name first, first name, middle name last and if you have any questions don't ask". We were being processed the same way and at the same time as the basic airmen, most right out of high school. Even though the room had many college graduates with all the OTS cadets, we not only all looked alike, we were all just as confused.

Everyone stripped down to their underwear and proceeded to go back through the same type physical we had received at the various induction centers. The major difference this time, was a battery of shots administered in both arms with big air pressure guns which made loud noises and hurt like the devil. There were a couple of guys who fainted and a few more who ended up with bloody arms from jerking as the shots were given. I was among those in the latter group. Next came our “ISSUE” of military clothes and stuff. We were loaded down with everything from combat boots to heavy overcoats. Most of the pants had to be altered for length and waist size. We were told to expect to receive them later in the day.

We were back at the barracks when my pants arrived. I did as I was told and tried them on. My two roommates went into fits of laughter as every pair of my pants from the dress blues to the fatigues were about 6 inches short in length and 4 inches big in the waist. How could they miss it that far? I soon found out, that in another squadron a similar thing was taking place only with the pants too long and the waist too small. After all my years in school with never having another student with the last name of Capps, I suddenly get to OTS and find I am not the only one. My OTS class was so large that several squadrons had been assigned to Lackland and several more were assigned to another area close by called Medina. I am not sure how they got us confused since other than having the same last name and same first initial, we had little in common. His name was Wyland Capps. He was short, stocky, married and came from Seattle, Washington. The mix-up of putting our "pants on" would not be near the problem which would come later of "pants coming off".

SETTLING IN

I have already mentioned that I ended up in the only room in our barracks where three people were assigned. One of my roommates was from the big island of Hawaii and he was a native Hawaiian. He looked like a movie star….tall, tan and handsome. I guess his tan was actually his native color but he had a face you would swear you had seen on a soap opera. His name was Don Kahana and he had attended military school somewhere on the mainland. He made it clear from day one, he was in the Air Force as a career and he planned to go after the highest officer trainee rank he could get.

In terms used today my other roommate would be described as a "NERD". His name was Leslie Eugene Francis, III and he was from Brooklyn, N.Y. Almost from the start, he took ribbing from the upper upperclassmen about his name. They would ask him if his name was Lesslie Francis or Francis Lesslie … "Which girls name is it"? He asked Don and I to call him Gene, which we did.

You could tell Gene had fought the battle of his name before. Growing up in Brooklyn , he was somewhat street smart. He had graduated at the top of his class from one of the Ivy League schools. I can't remember which one except I do recall him saying football was not something he cared for and he hated anything which had physical activity. In his case the "OBSTACLE COURSE" was properly named. He never did complete it and in fact he never passed any of the physical training tests. Even though he had one of the highest grade averages in the academic portion of OTS, he ended up going the route of self initiated elimination during the second half of our training. It was a shame to have gotten that far and then quit

The first morning we had inspection, Gene and I found out what kind of roommate Don was going to be. He did not help us clean anything. Other than make up his own bed and get his closet set up, he took off before inspection for one of his many interviews which were necessary to compete for a high OT rank. Gene and I collected two demerits each in our first room inspection. We both got one a piece because the overall room was not clean. Little Hitler had run his "white glove" around the light globe in the ceiling and also managed to pick up a smudge from the blinds. Gene got his second because his underwear was not properly rolled. I got my second because the laundry bag which we had to have tied to the end of our bunks did not have the correct number of loops on the string where you close it. (13 loops were required and I think I had 15). That one became my first idea of how to beat the system. By the next week before inspection, I tied it off permanently (perfect 13 loops) and had a zipper installed in the bottom of the bag. By the third week, everyone on our hall had copied my idea. I could have made a fortune if the cleaners which installed the zipper had paid me a commission on all the business I sent them.

At any rate Gene and I were only one demerit short of being gigged in for the weekend and we had only been through our first inspection. To say we were upset with our missing roommate was an understatement. He had gotten none since he was not there. Demerits were given out for everything from failure to salute or forgetting to wear your hat outside. Gene and my chances of making it the rest of the week without getting one more were pretty slim.

SQUARE MEAL & MY EXCUSE FOR LIVING

As I said the upper upperclassmen introduced us to the Mickey Mouse stuff right off the bat. We were to go nowhere alone, always two or more and of course we had to be in step. The first thing we were told to commit to memory was something called our "EXCUSE FOR LIVING". I never told

the upper upperclassmen but I had learned the little saying in a club (Ushers) I belonged to in high school. It went something like this: When asked "What is your excuse for living"? You were supposed to pop to attention and say
"SIR, IN AS MUCH AS EVERY LIVING CREATURE, NO MATTER HOW SMALL OR INSIGNIFICANT, HAS THE RIGHT TO STRIVE FOR ITS EXISTENCE … THEREFORE SIR, I TOO SIR, WOULD LIKE TO TRY TO LIVE. I THANK YOU SIR'. Since I already knew it, the first time one of the upper upperclassmen said it and said for us to learn it, I repeated it immediately (and said it real fast). They were all shocked and probably thought I was either a fast learner or had a photographic memory.

At our first meal we went through the chow line at the mess hall and everyone remarked at how great the food looked. We joked and laughed until we went to sit down. That's when little Hitler showed up and had us all sit at attention. Upon his command we were all to pick up our forks and eat or should I say attempt to eat a "square meal" I think maybe I had seen something like this in an old movie about West Point. The fork was to pick up a bite, come straight up, make a right angle toward the mouth, chew the food, drop straight down, make another 90 degree turn back to the plate. Needless to say, with the 15 to 20 minutes we were given to eat, we didn't have much of a meal. Luckily the upper upperclassmen tired of this game by the third day and we were given our meals in peace, I never did feel, however, that we had enough time. Every meal was rushed.

That first week was a blur. We had checked in on the 28th of December and then everything started happening so fast. If we had a New Year's Eve or New Year's Day I don't remember either. I cannot recall them being mentioned . I guess we just assumed if the Air Force wanted us to have those holidays they would have issued them.

GIGGED IN WEEKEND

I told you about the demerit system. Three demerits and you were automatically gigged in for the weekend. It was set up in such a way that everyone except those individuals who attained OT ranks of Captain and higher were stuck on base for that first weekend. Even a few of those who got ranks were given three or more demerits. I managed to get seven. What the hell, I figured after three it didn't matter. WRONG! What the upper upperclassmen failed to point out .... anything over six could be carried over to the next week. Like I suddenly find out I am only two away from my second weekend stuck on base.

Our Squadron Commander was an OT Major chosen from our class. He was the one who explained the carry over demerit system. He said that the reason it was stated that they "could" carry over was if you did something extra good during the week, it was possible to drop a few demerits. As our Commander, it would be his decision. He said, " I may as well tell you I do play favorites. Get on my good side and you come out ahead … get on my shit list and your ass is mine". He was about six foot four, from Mississippi and black. My future was looking bleak.

My jerk roommate from Hawaii had cost me a couple of my demerits since he was not around to help Gene and I clean the room. Of course his politicking paid off for him in a big way. With his military school background he managed to get promoted to next to the highest rank given to one of the officer trainees. He became an OT Lieutenant Colonel. By virtue of his new rank he was moved out of our barracks to be with the staff of the Commandant Of Troops, who was an OT Colonel. Our whole building had a shuffle with some moving out and some moving in. The three man room which we occupied was turned into a Squadron Staff Meeting room. Gene and I were both moved. I moved into the room which was next to the main barracks door with a guy from Ohio named Lester Coker. Lester had lost his roommate to an illness discovered during our Lackland physical. They never did tell us what it was. He was sent to the big hospital there and we were told that after he got better he would probably slide back to the next class. I did not know if I should try and sterilize his bunk but I was sure happy to be off of a top one.

Lester was a better roommate than either Gene or Don. He said right off, that we should each cooperate as much as we could to help each other and it would make getting thru the Mickey Mouse easier. Lester and I both had been given the OT rank of First Lieutenant. It turned out we were the only two on our hall. We had two OT Captains and of course the Squadron Commander (the OT Major, who had a private room at the far end of the hall). Everyone else held the rank of OT Second Lieutenant. The Squadron consisted of four Flights. Our hall or wing of the building made up one Flight and another wing of our building made up another. The other two Flights of our Squadron were in an adjoining building. Us having the Squadron Commander on our hall proved to be an advantage, although at first no one was particularly happy about it.

MISSISSIPPI'S MARVELOUS MARVIN

The OT Major's name was Marvin James Jefferson. His first order of business was to call a meeting (in my old room) of all those who held OT ranks of First Lieutenant and above. That meant Lester and I and the two OT Captains from our hall plus two other OT First Lieutenants and two OT

Captains from the other building. Major Jefferson, as he said he was to be called, wanted to deploy his troops in the best way possible to be sure each Flight had leaders living in close "proximity". I was trying to not form a prejudiced opinion but you've got to remember this was 1962 and it sure seemed this guy was trying to throw around big words to impress us. As it turned out the two OT Captains and two OT First Lieutenants from the other building already lived on separate halls. In our building one wing or hall had all OT Second Lieutenants. The Major asked that both OT Captains (they were roommates) to move to that hall and swap with two OT Second Lieutenants. I fully expected him to ask either Lester or me to also move over there but he didn't. He did ask each one of us to say where we were from and where we went to school. Other than the Major from Mississippi, I was the only other Southerner in the group.

SOUTHERN PRIDE

Major Jefferson and I hit it off in a strange sort of way at first. He started out in that first meeting using me as a sparring partner. In a joking way, he began a little game of using the fact we both came from the deep south to show everyone if a black from Mississippi could order around a white boy (and he did have fun with that term) from Alabama, anything was possible. At first I thought he was a smart ass and I was going to really be in for bad times ahead. After I got to know him and spent some one on one time talking to him, I caught on to what his idea was. He wanted me to take all his ribbing in a teasing way and dish it right back to him. He felt if we could joke about how bad racial tensions had been (and still were), it would put our squadron above all that and it would be easier for him to be accepted as the leader. He was right. It kept everything in a light, humorous vane and pointed out how ridiculous some prejudice can be. By the time we finished our three months our squadron won most of the top honors in everything from sports to drill competition, so as a team we functioned pretty well.

To give an example of how the major and I went back and forth with our kidding I can relate something which happened our second week. Each morning a different person was assigned the extra duty of running over to headquarters and picking up a copy of the weather report and uniform of the day instructions. This meant you had to get up at 5 a.m. instead of 5:30, which was early enough. Obviously if you did not guess the correct uniform of the day to wear over to headquarters, you had to change when you returned. And no, nothing as simple as a phone call could be used. The person assigned would stand at the end of the hall. As soon as reveille was sounded the weather report was read in a loud voice and the uniform of the day was announced. Due to strange fluctuations in the weather in San Antonio (especially during January and February) we often had to change uniforms several times a day but everyone all started out dressed the same. Major Jefferson and I talked ahead of time. I was assigned the duty on a Thursday and during the weather portion I threw in an extra comment. Since possible snow flurries had been mentioned with the coming of a Texas Norther, I said "I hope it snows us out of going to drill". I added "Wouldn't it be great if it snowed rear end deep to a 9 foot Indian". The Major's head popped out of his room at the end of the hall and he said "I might have known You would have a problem with an Indian. Maybe you need to do this again tomorrow For all you know my dad may be a 9 foot Indian". There were roars of laughter. The next morning I read the weather report just as it was written but when I got to the uniform of the day I said, "Uniform of the day … shower clogs, white sheets and don't forget your hoods". Amid much more laughter than the day before, Jefferson's head popped out and he said as he pulled back inside "Capps, boy, come on down to my room". By now everyone was out in the hall. As I got almost to the end of the barracks by the major's door, he stepped out wearing a sheet with a pillow case over his head. He said in a booming voice (as he lifted the sheet just enough for his bare feet to show), "Now when we go to the Klan

Rally y'all don't think anyone will notice do you"? He was quick and his wit and personality won him the respect of everyone in the squadron. No one ever knew that he and I had set it all up.

EXTRA DUTY ... VOLUNTEER FOR EVERYTHING

You always hear "Don't Volunteer For Anything". I am afraid that advice passed right over my head. I volunteered for everything which came along. As a result I had many experiences which most of my colleagues missed. Some were good, others not too great but my theory was if you volunteer for enough of the crappy jobs, when good ones come along you will receive more than your share. Also as an OT First Lieutenant, I was expected to take on extra duties. The Major and I had a good relationship. He found out that I used to do some sign lettering when I worked part time in college. I missed two days of drill doing signs he wanted for our squadron. I enjoyed getting out of drill so much that I decided to make it my goal to miss as many as possible. Remember I hated it in ROTC.

The funny part was I actually liked being the one who led the flight in drill. I was pretty good at that. If I was marching along side of the guys and barking out commands they had to stay in step with me. If I was in the flight with everyone else, I managed to step on heels, trip, turn right instead of left and all the usual screw ups. Our drill grade, however, was based on how well we led the flight when we were taking turns drilling them. I just did not like marching with the masses. At the end of each week all squadrons participated in a parade or Review (as they were called) They were highly competitive. As each squadron passed the reviewing stand they were graded. When we first started out, our squadron came in dead last. This did not sit well with Major Jefferson. He called all of his staff officers (those with an OT rank) in for a meeting. He asked for suggestions. I came up with what I thought was an ingenious idea. One that would not only get me out of drill but one which would make us the top squadron during a Review. One of the things I brought from home (which was still in my car) was a Polaroid Camera. I proposed that during drill each day, I stay up on the tower with the NCO who was responsible for teaching drill. I could take photos as our four flights passed and from them we could identify the guys who needed extra work. We pull all of them out and into the same flight and give them special attention. The ones who just could not stay in step even after the extra work we could send them to sick bay (headache, dizzy or whatever) just on the days we had Review and were graded. The Major thought my idea was brilliant. I hate to say it but on the way to my car to get my camera I got caught by a real Air Force Captain who gave me two demerits for not being with two or more other cadets. When he asked for my demerit slips (we were supposed to keep three on us at all times) I made another mistake. Since we changed uniforms so often I kept my slips in my hat. When I took it off to give them to the Captain, he added a third demerit for me being uncovered outside. Gigged in again. Our OT Major could do nothing to excuse demerits given by a real officer.

Back to the story about drill (we did it every day). First a note about our instructor. The description "Grizzly Ole Master Sergeant" fit him to a tee. At the squadrons first drill class he told us he had been in the Air Force for over twenty years and he thought most of us OTS boys were beyond hope. He had been picked to teach drill because at one time he taught drill with the joint military honor guard at The Tomb Of The Unknown Soldier at Arlington National Cemetery. At least that was his story. He made it sound rather impressive, pointing out all the Presidents, heads of foreign nations and various celebrities he had met during that tour. I'm not sure how much he exaggerated because he also said he knew who the unknown soldier was…...but was sworn to secrecy.

At our next drill class I climbed the tower and introduced myself. When I told him why I was on the tower with him and about my plan, he broke out laughing. He said if I was enough of a salesman to have talked my way out of drill and to have sold such a cockamamie idea I belonged up there with him, since he was the King of BS. I knew we were going to hit it off. After that first day we had already identified several guys in each flight who were consistently out of step. Gene, one of my original roommates was one of them. By our fourth day the sergeant and I had become friends. He was originally from Georgia and had been to Birmingham a couple of times. I think he even had a cousin who lived in Montgomery. After about a week it was almost like I belonged on the tower to keep him company. He started giving me advice on what to expect after getting my commission. He said the first thing that I should know is that NCO's run the Air Force. Get on the good side of your NCO's and they will take care of you. He gave me a lot of good suggestions which all proved to be useful.

One day he said "You do know that even though you are missing drill you will still have to be able to take command and drill one of the flights before you can graduate". He had been letting different guys try it at each of our sessions. So far he had only given around ten of them a final passing grade. I told him "I think I can do it right now, let me try it today". Believe it or not, in spite of skipping drill so much in college ROTC, I thought I had retained enough of the skill to fake it. I had a strong voice and could bark out commands loud enough to mimic the sergeant. I knew that the guys who he had already given grades to all sounded just like he did. He rattled off a series of commands and moves he wanted me to put the flight through. He then positioned himself where if I didn't give the commands properly they would march right into him. He loved it when I marched them to within three feet of where he was standing before reversing (to the rear march) or a couple times halting them within a foot. I was good at it and finally he said that I passed with flying colors.

Did my idea pay off? For every Review that followed we came in either first or second and ended up Honor Squadron for the last half of our training.

ESCAPE AT LAST

Our fourth weekend was approaching and much to my surprise the week leading up to it had gone rather smoothly. No demerits, and no problems. Friday was here and suddenly I was faced with having some time off and nothing planned. It did not take long for several of the guys to come up with an idea. As soon as we were dismissed around three in the afternoon, get the hell as far away from Lackland as we could. Actually more practical minds took over and a suggestion was made to go to one of the motels near the municipal airport where it was possible that some airline stewardesses might be staying. This was before they were called Flight Attendants, and as far as I can remember, they were all female.

Several phone calls resulted in two car loads (four in my car counting me) heading to a fairly nice motel not far from the airport. It was a typical motel of the early sixties, two floors, in a U shape with a small swimming pool in the center courtyard. We had been assured by a desk clerk that indeed, this was the place where flight crews had their lay-overs. Just the sound of those words, “lay-overs” had kind of a nice connotation and caused thoughts of fantasy which would have been worthy of an X-Rated movie (even though they were not rated back then). The guys in the other car arrived first and checked into two adjoining rooms next to the pool. We were only able to get two rooms two doors from each other on the second floor. We did have a great view of the pool and could see the other guys in their adjoining rooms directly across the courtyard. We just assumed the reason we could not also be on the first floor by the pool was those rooms were already filled with gorgeous airline stewardesses busily putting on bikini swimsuits and hurrying to go out and meet all the young, handsome, horny “Soon to be Air Force Officers”. Sure, dream on.
My roommate for the weekend, mainly because he happened to be riding shotgun in my car was a guy named Robert Mead from some little town in the Midwest. He decided he would do some

reconnaissance and check out the motel while I unpacked (which took two seconds). About the most important discovery he made was where the ice machine was located. Naturally the one on our floor was broken but he found one on the main level that had some ice. He did say that other than our guys from the base, all the rooms near the pool were occupied by families with small kids.

The two of us were out on the walkway in front our room when one of our guys from two doors down came walking up. We had been talking about how good it was just to be away from Lackland. He interrupted with what he said was good news. He had been talking to the guy at the front desk and found out four Western Airline Stews were on their way from the airport. He said he tipped the guy a couple of bucks to put them in the two rooms between us. The only problem was the two rooms which were connecting, were still occupied by a couple who had already checked out. The maid had told him she could not get in to clean until they got all their stuff out. The couple, who were in clothing sales, had stayed in one room while using the connecting room to show their line to prospective customers. I had been to several buyer shows like this when working at my parents store. The ingenuity of youth never ceases to amaze me. Before you could even blink there were four guys helping the couple load their car and carrying down trunk after trunk of clothing samples. We had them out of the rooms and on the road in less than ten minutes. The couple looked to be in their mid-sixties. I am sure they died many years ago but I bet for a long time they told the story about how four bald headed young men in San Antonio knocked on their door and told them they were there to help them load up because the rooms were urgently needed for some important Government "Undercover" Work. You cannot imagine how upset we were when the pilot from the Western flight took one of the rooms and the co-pilot took the other. The stewardesses somehow managed to get rooms directly below us on the ground floor facing the pool. The pilot turned out to be a former Air Force officer and took pity on us. We told him the story of how we had it all planned to get his flight crew in the rooms next to us, etc. He thought it was funny and gave us all of their names and suggested we might have some fun calling them on the phone. We were determined we were going to make contact before the other four guys who were now directly across from them on the pool level.

My plan was to call their room and ask to speak to one of them by name. I thought just telling them straight out who we were and how much trouble we had already invested in getting to meet them might work. I would suggest that we come down and introduce ourselves and then a couple of us go to a store and buy stuff to have a party out by the pool. The other guys quickly shot down my idea and said it would not work. The four OTS guys from across the pool would crash our party before it could even get started. I said okay I will go to my Plan “B”.

I had no thought whatsoever what Plan “B” was until I started dialing their room. One of the names the pilot had given us was Wanda Stearn. She was the Senior Stewardess. The reason I remember Wanda’s name was we started calling her Western Wanda almost immediately because of the airline. Anyway I asked to speak to Wanda and when she came on the line I introduced myself. I said “I represent the B.D.S.O.T.U.S.A.F. and your room has been selected at random from among guests in the motel to participate in a research project for the United States Government”. Hey, I figured the Government idea worked to get the couple out of their room, so maybe I was on to something. Wanda said “Who did you say you represent”? I had jotted down the initials as I started making the call. I repeated “The B.D.S.O.T.U.S.A.F. That stands for BLIND DATE SOCIETY OF THE UNITED STATES AIR FORCE. She started to laugh and I knew it was going to work. I told her we had been sent from Lackland Air Force Base as part of our training to become “Officers and Gentlemen by Act of Congress”. We were to locate four young ladies, introduce ourselves and then proceed to prove that we could indeed be gentlemen and show them a good time. I even threw in that our research project had received the endorsement of the San Antonio Chamber of Commerce. I heard muffled talking and laughter and then Wanda came back on the phone. “Where are we going, what should we wear and how soon do you want us to be ready”? BINGO!

I told her we would work out the details and call them back but whatever we did, the dress would be casual. She made some smart remark about us wearing our uniforms. I countered by saying “We will if y’all will”. We did not even bring our uniforms but I knew they did not want to wear their stewardess stuff any more than we wanted our Air Force junk. We did have a major obstacle to overcome. Now there would be eight of us. My Comet convertible would be a bit crowded. After a little discussion, we decided it would be fun to go to a good Mexican restaurant. If we had to get a cab for our overflow, so be it. Just as we were about to call the desk clerk (who we now thought owed us one) there was a knock on our door. Two of the guys from the rooms downstairs came to excitedly tell us that there were two rooms full of airline stews directly across from them. You can imagine how cool we thought we were when we said “ Yeah we know, we have dates with them for dinner”.

As the two guys started back downstairs the airline pilot walked by our door and asked how we made out calling his crew. We told him what we had done and asked if he knew of a good restaurant. He said as a matter of fact he did. There was a great Mexican place which even had music. The food was typical Tex-Mex and the prices were reasonable. Sounded great to us. Wanda turned out to be the oldest (26, which was 4 years older than I) and to be honest the least good looking of the four. Since I had made the original contact and she had been their spokesperson we sorta ended up together. She solved the space problem by saying to two of her cohorts "Virginia and I will ride with Bill and Robert and y'all four (pointing to the other two guys) take a cab and meet us at the restaurant". "Wanda was a take charge lady. She had been to the Mexican restaurant before. I found out later she had been there with the pilot. She told me that he was married but liked to play around. She had a brief affair with him but decided he was trouble with a capital "T" and now they avoided each other when they ended up on the same flights.

I decided Wanda's telling me all the personal stuff so quick was her way of letting me know she was experienced and had flexible moral standards. In short I could be in for a weekend to remember. Funny how the dreams and fantasy always beat reality. Wanda was loud and after three or four margaritas rather obnoxious. I had maybe one or two. Back then there was not the pressure of today of having a non drinking designated driver but I was always responsible enough to know that moderation would keep me out of trouble. Before long everyone but me seemed to have had too much. The co-pilot and a Mexican girl walked in the restaurant. Wanda yelled at them to join our table (which by now was loud). He did not act like he was too happy to have been recognized but they joined us anyway. The time flew by pretty quick or maybe we started out later than I remember but it was near one a.m. when we headed back to the motel. The co-pilot and his date had left earlier. That left eight of us who somehow, with the top down, managed to get in or on my car. Remember it was a Comet convertible with a 4-in-the floor stick shift. Robert and his date (no longer Virginia, not sure when that swap took place) were sitting shotgun. Wanda was somehow sitting on the console, making shifting gears an experience I have trouble explaining. It sounds better than it was and makes a great story but with the other four in the back seat sitting practically on the trunk, it was an impossible situation. And it was one which attracted the attention of a pair of San Antonio's finest. I noticed the red light in my rear view mirror. Yes, in '63 they were red and not blue.

Since I was the driver and the only one not singing 99 bottles of tequila on the wall, I got out to talk with the police. They said we were not doing anything wrong, other than being too noisy and the four people in the back seat would definitely have to not sit up on the back while the car was in motion. There were no seat belt laws back then. That crazy group never quit singing the whole time I talked to the officers. The policemen guessed we were from the base (our haircuts always gave us away), but when they heard we were in officer training school, they just said be careful and hold it down a bit. I would love to report that the rest of the night ended up as a wild orgy but other than the co-pilot and his friend ending up skinny dipping in the pool at 3 a.m. it was tame. Of course by the time we got back to the base it sounded like we had the wildest weekend of our lives. Everyone embellished the story with a twist of their own and possibly the stews told it differently on their next flight. I swear I read something which sounded vaguely familiar years later in a book called "*COFFEE , TEA OR ME*", written by two stewardesses.

WHAT DID YOU SAY YOUR NAME WAS?

Even though we rarely had a chance to enjoy it, we did have our own Officer's Club …. or should I say Officer Trainee's Club. We were not allowed in the real Officer's Club (which was pretty nice).

Our club was only open Friday (after noon) and Saturday and Sunday. After the first couple of weekends where everyone headed to a motel to "escape" Lackland, the reality of not being able to afford that kind of expense set in. Most all trainees headed to our own little club. Actually there were two, one at Lackland and the other at nearby Medina. That's where our overflow and the few WAFS going thru OTS were located. It was on one of those "head to the club" weekends when I finally almost met the other Capps. We went to the club at Medina because all we had was a ping pong table, a pool table and a bar ……. but no women. At least Medina had the WAFS. Remember the other Capps was the one where our clothes got switched.

The WAF officer trainees were not hard to spot. When Robert and I walked in we saw that each group which had a couple of females also had a crowd of eager males surrounding them. We tried to mingle as best we could and finally managed to work our way into a conversation with two of the WAF trainees. When we occasionally saw the women at our parades they were wearing their uniforms, and they were not that attractive as a group. Now in this setting in civilian clothes (especially after isolation from the opposite sex all week) they began to look pretty good. One of the two we were talking to was very attractive. She was about five foot four, short blond hair, a beautiful tan and a body worthy of a Playboy centerfold. They were holding court with about ten guys (us among them) hanging on every word. The other girl was not as attractive but she had a super laugh and she was wearing a low cut dress which drew the attention of most in the circle. Everyone introduced themselves and made small talk about the weeks events and our training.

I was reasonably sure the blond did not hear my name because she kept talking to a tall handsome trainee (who was wearing his Dress Blue uniform). I always hated those guys. They were the ones who looked like the men you saw on Recruiting Posters so they wore their uniforms everywhere. I could not wait to get in civilian clothes anytime I got the chance but then again I did not look like one of those poster types. The girl with the funny laugh gave me a real strange look when I said my name. I saw her whisper something to the blond when there was a break in conversation. The blond looked straight at me and said "what the hell did you say your name was"? When I said "Bill Capps" she blurted out "Then who is that guy over by the bar"? She pointed to a short fellow ordering a drink. One of the men standing in our group said "I think his name is Wyland Capps". With that the blond charged over toward the guy and a loud argument erupted with the blond using language probably not appropriate for future female Air Force Officers.

The remaining girl put her arm around me, taking me off to the side she said “I believe I can tell you what is going on”. Apparently Wyland who was married and had two kids (we found this out later) had gone off base the weekend before with the blond. They drank a lot and it ended with them shacked up in a motel. It was there he registered as Bill Capps, which was the name he had also given her. Wyland apparently decided that if our pants could get mixed up going on, they could also get mixed up coming off. I never did get to meet Wyland or especially the blond. I wanted to find out if I had a good time and if she had given me a good reference.

TRICKS OF THE TRADE

It is not hard for me to understand how captivity brings out the genius in prisoners. Necessity forces you to be creative, cunning and at times downright deceptive. In some of our classes we were shown films of what to do if captured by the enemy during wartime. We saw interviews with former prisoners of war from World War II and Korea. We also saw captured films the Japanese made of them actually torturing our prisoners. They were brutal. In a sense we all felt like we were all prisoners when we went to the barracks at night. We would watch those training films and then imagine what it was like to endure that kind of treatment. I guess it meant to show us what we were going through was nothing. We still felt like we were having torture of our own. We had few privileges and we were constantly being harassed both by real officers and OT upperclassmen.

The human mind can come up with solutions when challenged. Once we were back in our barracks, after lights out, we were not supposed to be up studying. A large majority (me included) had to struggle with the academic portion of our training. The trainees always found ways. On any given night at any given hour you could discover cadets under a blanket or under their bunks with a flashlight studying for a test. Even closets were popular studying spots and before tests every commode in every latrine would be occupied.

The tricks used for study were nothing compared to the hiding places cadets found (or in some cases constructed) to conceal junk food. Candy bars were NO NOS in the barracks but they were easily available on trips to the BX and off base on weekends. The problem was where do you stash them so they are available when you are not supposed to have them. Somehow they always tasted better then. Hiding them became an art form. Upperclassmen took pride in finding stuff (which they confiscated) during inspections. We looked forward to becoming upper upperclassmen because we knew to the victor belongs the spoils.

Yep, the plundering starts in training long before you see the real thing in combat. Common hiding places ranged from removable vent covers, bed post tops and even in laundry bags. You've got to be pretty desperate to keep candy in your dirty clothes. One guy even went as far as cutting out a rectangular opening in the pages of an old Bible. Nothing was sacred in OTS. Unfortunately he was one of the few cadets who even had a Bible and his secret got out .... when during an orientation tour a visiting Base Chaplin asked to see it.

I devised a hiding place which was never discovered. In fact I was so proud of it that I passed it on to the cadet who I had to sponsor in the class behind mine. Each cadet was allowed to have one box of laundry detergent in his locker. On one of my weekends off base I purchased a large box of Tide. In fact I bought two. I poured all of the contents of each into a bag. With the aid of only a single edge razor blade and a roll of Scotch tape I made a false bottom about two thirds of the way down. I used the bottom one third of the second box as a tray which the false bottom box would slide over. The top part of the box was then refilled. I could hide so many candy bars and junk snacks in there that I soon started selling them to my fellow classmates. They could never figure where I kept coming up with all the merchandise but with supply and demand I managed to not only make many new friends but also a nice profi

DESPERATE TO MEET THE OPPOSITE SEX

Rumors about saltpeter being used in the food to curb the sexual appetites of OTS students were common. A guy in our squadron who's dad was a doctor said that it was just folklore and might cause a headache but did not reduce libido. That made sense because as every weekend approached you would hear guys plotting how to search out members of the opposite sex. It was tougher than you think. Most local females wanted nothing to do with anyone from the base. Of course the local prostitutes were the exception. I am talking about meeting nice girls even though I guess most everyone dreamed of those who were nice but not too nice. Most of the San Antonio women were used to the fact that guys from the base were only going to be around for a short period of time and they would rather not get involved. We thought that since we were in school to become officers that it should give us an advantage over those guys only there for basic training. WRONG ….. we had shaved heads the same as they did. We all looked alike and they were just as on the make as we were. The only slight advantage was that many of us had cars and a bit more disposable income. The problem, however, was the same. There were not that many women and you had to come up with an original idea in order to meet them.

THE LADY OF THE LAKE

After our weekend with the stewardesses, I sort of became the social chairman of our little group. Even though it was only somewhat successful, I had come up with a clever way to get us introductions. I was appointed to find a way to meet some women who were "nice". My instructions were to find some females who were not in the business to separate us from our pay checks. Thinking back, there were quite a few working girls in San Antonio back in early '63. My plan was simple. There had to be some place in the city where there were a bunch of women who would not mind meeting some future Air Force Officers. I went through the phone book looking for nursing schools or wherever I might come up with a dorm full of women. I ran across "The Lady of The Lake Catholic Women's College". When I was in Auburn I used to be in charge of pledge swaps between my fraternity and different sororities. I figured why not employ that strategy now. I called the college and told the woman who answered who I was and that I needed to talk to someone in charge of the social functions for their college students. I explained that I had been asked (I never said who asked) to set up a get-together for a group of officer trainees as part of their training. It was to teach them how to handle themselves at Air Force social functions. I said

"We of course, will handle all planning, decorations, refreshments and expenses. Your students will be invited guests". I was thinking damn I'm good. There was a pause, then I could hear her talking to someone and when she she came back to the phone she said" Give me your name and number and I will have the Mother Superior call you back". I gave her my name and the pay phone number.

We sat around that phone booth for almost thirty minutes. The guys started giving me ribbing saying this hair brained scheme won't work but then the phone rang. I had one of the guys answer and say "Lackland Air Force base, Officer Training School, how may I direct your call"? The Nun on the other end of the line asked for me and she was told "Just a minute please". I waited a few seconds before I picked up. I went through the same spiel I had used when I talked with the lady who originally answered.. The Mother Superior had an intimidating voice and I almost chickened out in the middle of my proposal but I thought, all she can say is no. Much to my surprise she said "That is the greatest idea I have heard. General Beaman out there is one of our school's big contributors, did he suggest that you call me"? Uh Oh, I may be in trouble again. "No ma'am, I'm not sure who told me to call but we would be sure to include the General on the guest list". That did it, she said "When can you come out so we can discuss this.

We would love to have the party here at the college. We have a lovely recreation room in our administration building". Unfortunately, since we could only get out on weekends my plan now appeared to be at least a two weekend deal. I made the appointment for Friday afternoon. I conned my roommate into going with me so I would not have to face the Mother Superior alone. The college was beautiful. It was small but The Lady Of The Lake College indeed had a lake. On the way over we ran thru all the Nun jokes from "Any friend of Zorro is a friend of mine" to "Try not to get into any bad Habits". We decided we had better be on our best behavior if we were going to pull this off. Fortunately for us the Mother Superior had a crisis that needed her attention. After greeting us and exclaiming how nice we looked (we had worn our dress blue uniforms and our spit shined shoes), she introduced us to a young attractive Nun named Sister Veronica. She said we could work out all the details with her. Sister Veronica had not been out of the convent long and it did not take her three minutes to put two and two together and figure that the Air Force had absolutely nothing to do with this idea …. but she thought it was fantastic. We leveled with her and told her we thought this was the only way we could think of to meet some nice girls.

Sister Veronica sort of took over. She said if we could handle inviting about 30 guys she thought she could at least find that many students who would want to attend. We would set it up for the next Friday night. Dress would be casual. We would serve punch and maybe have small sandwiches and cookies. They had a record player and the girls could bring records in case anyone wanted to dance. She said if I could collect about $5 from each of the guys that would be enough to cover everything and she would get the students to prepare all of it. Back in 1963 one hundred and fifty dollars went a long way. This was beginning to take shape. I asked what do we do about inviting the General since I had said that to the Mother Superior. She said "She won't remember it but it would probably be good politics for the school to drop him a note and say they were hosting a social function for some of the Officer Trainees and say he was welcome to come". That sounded okay to me. He probably could care less but at least the college would remind him they were still there and needed his donations. Besides it would give some legitimacy to what I had said to the Mother Superior.

When Sister Veronica found out I was from Alabama and had graduated from Auburn, she said "I knew I liked you, I'm from Mobile". She said "We have a student here from Tennessee, I think I will call her to come down and help us. I believe the two of you will have a lot in common. Her brother graduated from Auburn". When Linda

Haley walked in, she lit up the room. She had the kind of personality that made you instantly like her. With a warm friendly smile she bounced in and asked Sister Veronica "What's up, you getting arrested? Oh these guys are not cops are they"? I decided right off the bat, I was going to have a date with her for the party. We went through normal introductions and finally got around to talking about the next Friday night. Linda said she would be happy to help set everything up and I could work with her by phone to get things coordinated. We left the college feeling pretty smug that we had pulled one off. We were headed downtown when it suddenly hit me. Why didn't I ask Linda if she had plans for that weekend. Here, I finally meet someone and I let the opportunity slip by. During the talk about the party, I hadn't even gotten to ask about her brother who had gone to Auburn. Who knows, maybe I even knew him although I could not think of anyone named Haley.

I pulled into a filling station and called her. Lester was now coaching me saying "Ask about tonight and see if she has a friend". I told her my roommate and I got to talking on the way to town and we just wondered if she might like to go out and was it possible she had a friend. I guess I was talking pretty fast. She said "As a matter of fact the answer to both questions is YES" My pulse quicken a bit until she continued. " Yes I have a friend and yes I want to go out .... and he is picking me up in just about half an hour". She knew she was having a little fun at my expense but I tried not to show the embarrassment and my disappointment in my voice. I said "Well I knew anyone as cute as you would probably be going steady but I thought so far this has been a fairly lucky day so why not give it a shot. All you could say was No". She said "Excuse me, did you hear me say No? I believe I said the answer to both questions was Yes. My friend is just a friend and if I call and tell him I have a date he won't have a problem at all". We went back to pick up Linda and her roommate at 7 (which gave us time to get back to the base and out of those stupid dress blue uniforms).

THE PARTY

My double date with Linda and our respective roommates on the Friday preceding our planned social function turned out great. In fact Linda and I went out again (just the two of us) on Saturday night and also got together for a picnic on Sunday afternoon. So as I had hoped when I first met her, I had a date with her for our party. The two of us talked almost every night trying to set up plans for how many guys were coming, refreshments, etc. We collected from 32 guys but only 27 actually made it. We had posted a map but several said later they got lost. I guess the idea of a "PARTY WITH OFFICER'S TO BE" (which was the sign Linda posted) must have appealed to the Lady Of The Lake students because around 40 girls showed up as well as 4 Nuns, a priest and 2 maintenance men who decided to stay for sandwiches. The General had called the Mother Superior to tell her what a great idea the party was and to send his regrets. She stopped by to relay the message but did not stay. There was plenty of food and the consumption of punch went up considerably when someone added a fifth of vodka. I was not around but was told that the Priest not only approved but he was the one who originally suggested that "It needed something".

To say the party was a success would be an understatement. The girls outnumbered the guys, so they became more aggressive and asked the guys to dance. Even the shy guys were surrounded by females and conversations were easy to start. From that night until our class graduated there were many dates which were set up from that original "Social Function".

FIELD DAY AND THE OBSTACLE COURSE

As graduation neared we began to take final exams in all our academic subjects. In addition our physical program started to get tougher. We had tests in everything from sit-ups to pull-ups, to running the mile. A test known as the Five BX training program was big back then and every officer on active duty was required to pass it each year. We were all in such great shape because of the stuff we had to do, the Five BX was like a walk in the park. The tests for OTS were ten times harder. I had never been in as great a shape as I was then nor have I been since.

All the physical stuff culminated with a Field Day Competition between all squadrons. Our squadron field day team was led by none other than our own Major Jefferson (you remember, Major Marvin James Jefferson). By now most of us had nicknamed him "M.J." (that was many , many years before Michael Jordan). The Major said we could call him that as long as the "M" stood for Major (he hated the name Marvin). Well M.J. was by far the best athlete in the squadron. He had played football and run track in college so he was in great shape when he arrived on base. He paced our selected teams in the field day competition and we won going away. There was no squadron even close. Second, third and forth were only a couple points different but we had the meet locked up before the final event which was the mile relay. I mention it because it was the only thing I participated in for Field Day. The two best people were selected to represent the squadron in each event and I was usually third or forth. M.J finished First in the 100 yard dash. First in the 220, and he was to run the anchor leg in the mile relay. I ran the third spot which meant I would hand the baton off to him. The pressure was off because we could have finished last in the relay and still win the meet. M.J. gave his pep talk before the race and it was obvious he planned for us to win this event too. He told me "You drop that baton and your white butt will be in deep shit". We were in close second when I passed the baton to M.J. but he pulled away and won by five yards.

After the awards and we got back to the barracks, M.J. asked me to come down to his room. Even though my showing in the Field day Competition would not make any highlight reel, he wanted me to know he appreciated my contribution and my whole attitude in general during our OTS training. He said "You and I worked well together and I think you helped make our squadron what it became. I know I got most of the credit but I am trying to personally thank all those of you who helped me". That meant a lot. Then he said "I've got some "Inside Tips" for running the Obstacle Course that I am passing to only a select few in our squadron". We were set to run the Obstacle Course the next morning. All during our training it had gained a reputation. Many of the cadets dreaded having to run it. We had heard bits and pieces about it and they were all bad. It was the same one the Basic Airmen were required to run …….with one exception. On the day when the OTS cadets were to run it, all of the NCO's and most of the Officers who were instructors turned out in force

…. to laugh and make the course as tough as possible. You've seen obstacle courses. Each person starts at the same point and runs against time over a measured course. You swing by ropes over ponds, walk hand over hand on ladders across creeks or mud holes or both, climb walls with cargo nets, scale cliffs with ropes … you get the picture. The NCO's delight in the start where you must dive out and catch a swinging rope to land over a giant mud hole. You miss that first obstacle and the rest of the course is miserable. It is bad enough to negotiate the trails and other obstacles dry but being wet and covered in mud can make it most unpleasant.

M.J. gave me some good tips which he had received from the NCO who helped set up the course for the OTS run. I improvised a few on my own but here is an account of the way I remember running the course: First you must run it wearing your fatigue uniform and combat boots. To keep from ruining the spit shine (which I spent about a month putting on those boots), I covered the toes with electrician tape. This kept them from being scuffed while crawling through the dirt under barbed wire. Since most of the time I was running, no one ever noticed. The start was indeed over water and a big mud hole. An NCO would swing the rope from the other side. You had to dive and catch it with enough momentum to carry you far enough to clear the mud pond. I had been warned

“Anytime you see a larger than usual group of observers at any point ….. speed up do not hesitate. They are there because that’s a spot where OTS cadets bust their ass”. The first pond was a good example of a crowd. I noticed that the NCO would swing the rope where you would naturally grab it too low. When you went to swing, your feet would drag the water and it would cause you to fall off into the mud.

I never slowed down. When the timer said GO, I broke into a dead run, took a huge leap and could have cleared the pond without the rope.. I did grab the rope high and with a short swing cleared the other side by three feet and never broke stride. On the trail I passed as many people as I could. Everyone ran against their own time. M.J had said if you can get ahead of the masses it will be easier when you get to the ladders across streams. They won’t be wet if you pass the guys who fell in the mud. When I came to the second ladder I hit a problem. An NCO was making a cadet who fell in the water climb back up and try again. I actually passed him by going hand over hand on the sides of the ladder. The NCO yelled at me but he was on the other side so I never stopped.

I ran until I came to the TEAR GAS HOUSE. An NCO was at the door wearing a mask. Each cadet was handed a mask, told to wear it inside the house and when the order was given, remove the mask, deposit it in a basket and find your way out the back of the house. I never broke stride, grabbed a mask, pulled it over my face and dashed inside. The house was full of gas but I could make out the image of an NCO standing in the center of the room pointing in the direction we were supposed to go. I caught a glimpse of a flash of light toward the rear of the building. I realized there was a window which had a canvas flap hanging over it on a back wall. When the order came "REMOVE MASKS", I saw the flap go up. I yanked off the mask, dropped it in a basket and ran and dove head first out the opening. I landed on top of whoever was going out ahead of me. I guess I instinctively tucked and rolled because I bounced off the guy and came up on my feet, and never looked back.

I ran down a trail until I came to the base of a cliff where 4 ropes were hanging down from the top. We were to choose a rope and climb up the cliff. I did not like the fact that several people were climbing the rope at the same time. If the person or persons above you were slow climbers, you had no choice but to go slow too. I noticed the rope on the far end was a bit higher than the other three and since there was only one other person on it and he was near the top, it was the one I chose. I scampered up pretty quickly, easily passing those on the other ropes. As I neared the top, a guy on the next rope over was obviously having trouble climbing. He yelled to the NCO at the top "I can't make it". Just as I swung my legs up on top, the guy just let go of the rope and tried to jump back down. Now this was not a real high cliff but it was over 20 feet. When he jumped it was about the equivalent of a 1 and ½ story building. He hit two people on the rope below him. One managed to hold on but the other fell with him. He was not hurt but the guy who jumped I found out later, broke his leg. The NCO at the top told me to head out so I kept on running. I am sure this accident slowed those behind me and gave me an edge on the others in my squadron (the fellow who jumped was from another group).

I breezed through most of the other obstacles because no one was near me. Just before the final dash I came to the live fire field (they tell you it is live ammo so you won't raise up). M.J. had said they were blanks but I was not going to test it. This portion was a maze of trenches and barbed wire which you crawl through while guns are supposedly firing over your head. Mortars are exploding on all sides. Since by now I was so far ahead of everyone else, I found myself crawling thru the maze alone. I guess they didn't want to waste too much firepower on one person because when I was going under the wire on my back only two machine guns went off with short bursts and they blasted only one mortar. When the rest of my squadron went through they said it sounded like World War III. The last ½ mile was an all out sprint and I did did it as fast as I could. When I came to the NCO who was the end timer, he actually questioned if I was wrong about my group and their starting times. I was not only the first to finish from my squadron but I was ahead of several other squadrons as well. As a reward I was allowed to ride back to the barracks in a staff car. The NCO also remarked about my uniform looking clean and the fact my boots were not even scuffed. He laughed when I started pealing the tape off the toes. I enjoyed running the course so much that I was going back the next day with M.J and two other guys to run it again. We were going to try and break the record time. I had done it pretty fast but when we found out I was not even close to the record we all decided once was enough. I think M.J. was a little pissed that I did beat his time.

FINALS WEEK ....THE END IS IN SIGHT

I crammed for the academic portion of our tests the same as we had done when we were in college. As usual some people found them easier than others. I think I mentioned before that I have never done well with true / false or multiple choice so I burned the midnight oil quite a bit. I was not sure we would actually receive grades ..... I thought it was more like pass or fail. In writing this book in some old files, I came across what appears to be my grades. I averaged an 82 on everything academic and graded 96 in the physical training. I do vaguely remember that Don Kahana was honored for being top in our class but then again he came from a military school. Even though he held the second highest OT rank of Lieutenant Colonel I had stayed in contact with him during our training. When M.J. found out that I knew him, he had me use that original roommate connection to get us a few perks for our squadron. I could walk into Command Headquarters and see Don even easier than Major Jefferson (and with him being a Squadron Commander, he reported to Kanana).

My memory of that last week has somewhat faded. I do remember being worn out, both with staying up cramming with no sleep and also with the excitement of the unknown. Like what would lie ahead after being declared an Officer and Gentleman by an Act Of Congress. The tension was mounting as to what our first assignments would be. The guys who were going to pilot training, all pretty much knew where they would be going. It was guys like me who would be classified as Administrative Officers who had no clue where we would be sent. We did get some good news. One of the real Air Force officers assigned to OTS called all the soon to be Administrative Officers together and said "Well you guys have it made. You will all stay in Air Training Command but you get to pick a choice of three bases and you'll get assigned to one of the three. He gave us a list of all the ATC bases. I was happy to hear that. I tried to pick three which were close to my hometown of Birmingham. My first choice was Kessler, AFB in Biloxi, Mississippi, my second was Maxwell, AFB in Montgomery, Alabama ,and my third was an ATC base in Georgia. I could not wait to call my parents. They were making plans to fly out for the graduation ceremony and now I could tell them to only buy a one-way ticket because I could drive them back to Birmingham on my way to my first assignment.

FIRST RULE OF OTS .... BE FLEXIBLE

The whole final week became a blur. We were expecting good weather for the graduation ceremony so a Review (Parade) was planned. Also there were rumors that we would have a fly over of jets at the moment we were to pin on our Second Lieutenant Bars. Of course we were told NOT to throw our hats in the air like you see at all service academy graduations. It was also added, if you do, nothing will happen accept you might lose your hat. Tradition has it that you get an NCO to pin on your gold bars for luck and also for luck you need to give the NCO a dollar. No doubt the tradition was invented over a few beers one night at the NCO Club. My buddy from the drill tower and I had already planned for him to pin on my bars and as many others from my squadron as possible. My parents plans were made, They would fly in the night before graduation and as soon as it was over we would take off in my car to Birmingham (by way of one night in New Orleans to celebrate). This made a lot of sense because I would be heading to one of the three Air Training Command bases I had chosen, all of which were East toward Birmingham.

Guess again. The Air Force wanted to keep us off balance (and their favorite word) flexible. When the first assignments were posted, guys were running around all excited about where they were going. I didn't even fight my way to the bulletin board because I thought no matter which one of the three choices I picked, I would be happy. One of the guys who's last name was Barrett walked by and said "Well looks like you and I are going to Amarillo". What? Amarillo, that wasn't possible. It was not one of my choices. Hell, I didn't even know where it was. They must have made a mistake. I fought my way to the front of the crowd which was still pushing and shoving. Sure enough .... there it was on the bulletin board. All the Administrative Officers in alphabetical order. The first five, me being the last, were all assigned to Amarillo Air Force Base, Amarillo, Texas. I would pull my first tour of duty and not get out of the state of Texas. Of course we all felt an error had occurred but we soon found out. Better get used to it "That's how the Air Force does things".To say I was disappointed would have been an understatement. To say I was mad would be more accurate. I never had a chance to complain. We were told in no uncertain terms that assignments are final they would not be changed under any circumstances. Of course a few months later when I asked why Barrett never showed up in Amarillo I found out he somehow ended up in Bermuda. Just a coincidence I am sure but we heard his dad worked in Washington.

LAST NIGHT FOR A FLING

Graduation was set for Saturday afternoon. For my parents, I had made reservations for Friday night at a motel not far from the base. Actually it was more like the old time tourist courts with the rooms being in separate little houses. It was not fancy but all I could find as rooms nearby booked up fast. I found out my parents flight would not arrive until late in the evening. Seemed a shame to pay for a room which would only be used for a short night.. The tourist court where I had a reservation did not have a pool but it did have a big sign which read "TV IN EVERY ROOM" (not all motels back then had them …. at least not in EVERY Room). Some of the guys in my squadron found out I had rented a room. They suggested we go use it since my parents were getting in late. I had gone over early to try and see how it looked and since it was already empty and cleaned they had given me the key. Four of us went over in two cars around 10 in the morning. We had no idea what we were going to do but felt we had to do something to say we were celebrating. We were not alone in wanting to do something to celebrate graduation. Many guys were planning parties. My old roommate, Don Kahana and several of the higher ranking OT Officers had splurged. He had told me that they rented rooms for Thursday and Friday nights at the Davy Crockett Hotel by the Alamo. I was familiar with it because I had stayed there the two nights before I checked in at Lackland. Four of the officers had gone in together and rented two adjoining luxury rooms which opened on to the pool. A little (no, much) more expensive than the one I had stayed in.

Maybe we had been in my rented room about thirty minutes when it became obvious all we were going to do was sit there and drink. The guys had beer and a fifth of vodka and maybe some bourbon. When I told them about Kahana they all wanted to know how they got off base on Thursday. Apparently even OT "Rank Has Its Privilege". Everyone had their swim suits except me. I forgot to tell them my motel had no pool. Someone said "Hey let's go crash Kahana's party, we can use their pool. We had all decided that with graduation the next day their OT rank meant nothing now …. we were all going to wear the same gold bar. What we did not know is they had partied until the wee hours of the morning.

We arrived unannounced around 11 a.m. The desk clerk had given us the room number (they did that back in the old days). Knocking on the door brought no response and we could not hear anything stirring inside the room. Maybe they were at the pool. Before we went to try and find it, one of the guys banged hard on the door and said "OPEN UP …….. THIS IS HOTEL SECURITY"! We heard scrambling and it sounded like things getting knocked over and a nervous voice relayed "Just a minute, I'll be right there". When the door opened a crack, we pushed our way into the room yelling and laughing. Kahana had a sheet wrapped around him and he looked like death warmed over. A Mexican hooker was scrambling to put on her clothes. One of Kahana's high ranking friends was out like a light lying face down, butt naked on the floor. Another guy had made a quick exit out a sliding glass door and was trying to walk away nonchalantly. His stagger gave him away but he was determined to distance himself from whatever was coming in the room he had just fled. Clothes and bottles were everywhere. The now almost dressed Mexican headed out the sliding door which was still open. The room was a total wreck. Kahana stammered something which was either what the hell were we doing there or maybe it was get the hell out. Whatever he screamed caused two nude girls in the bath room to bolt out yelling and cussing in Spanish. The scene was straight out of one of those porn movies …. a bad porn movie.

The one thing I most remember was Kahana saw that one of the girls (saying girl was generous by several years)… anyway one of the ladies of the night had his wallet in plain view. Well she certainly didn't have a place to hide it. Kahana yelled "Grab my billfold, don't let her get away with my money"! He dove for her hand but she threw the wallet to the other girl. Kahana got tangled in his sheet and fell on top of the guy passed out on the floor. Now as hotel rooms go, this one was large but not near big enough for all this activity. There were four of us and four of them. The next few seconds were a blur of boobs and butts. Somehow they found enough of their clothes and came to a mutual agreement on the wad of money that now the older of the two ladies had clutched in her hand. Kahana got his wallet back but not without having apparently paid for his evening and that of his friends as well. With money in hand the two exited out that same sliding door to the raised eyebrows of an older couple who had witnessed most of the comedy while sitting in lounge chairs by the pool. If there was a fourth guy in the connecting room, we never saw him. We left as Kahana and the guy from the floor said they needed to hurry and get dressed, check out and get to the airport. I hope old Don boy did not have to explain how he got sets of fingernail marks on each side of his back. Of course remember they were staying next to the Alamo …. A Bowie knife could be blamed.

It was still early so everyone wanted to go back to the base. I decided to try and set up a last date with Linda. I had mentioned to her earlier in the week that maybe we could get together to celebrate my last night in San Antonio. Every since the party at The lady Of The Lake, she had been the only girl I had dated. We had not been out lately because I had been busy trying to graduate and she had been tied up in school. I think we both liked each other but neither of us expected our relationship to proceed to a serious one. It had not all been platonic. There were several hot necking sessions, but always stopping short of going too far. Linda had said that by Lady Of The lake standards we should probably already be engaged. I picked her up around 5 p.m. We went out to dinner and generally made small talk avoiding discussing the fact that after graduation I was probably gone forever. The subject of me renting a motel room for my parents who would be coming in later did come up. I said it was a dump but all I could find. I told her that a group of guys and I had used it earlier to celebrate but since we did not want to just stay there and drink we had come up with an alternate plan. I then told her about what we discovered at the Davey Crockett. She laughed and said "I am glad you did not celebrate the way they did'. And then she added "You were not planning on taking me to your motel were you"? I smiled and said "I would be lying to say the thought did not cross my mind". Then she said "I want to go see it ..... just to see if it really is the dump you said".

Here is where the plot thickens. I said I hope the key is where it is supposed to be. Somewhat confused she said "Why don't you have a key"? I then explained that one of the guys with me earlier had forgotten a wind breaker and said he and a couple of guys were going off base to pick up uniforms they had altered and wanted to stop by and pick up his jacket. Since I would be leaving to pick Linda up I just told him to drop the key back at the desk. I said I hope he remembered to do it. We arrived around 8 p.m. and she agreed from the outside it did look like a dump. The key was at the desk.

When I opened the door I was hit with a shock. The room was a total wreck. Apparently those guys had stayed and had a late afternoon party. There were beer cans and cigarette butts everywhere, The place smelled of smoke and stale beer. The bed had a combination of popcorn, chips and mud. Visions of me bringing my parents here later were not good. Linda laughed and said "Now I know why you really wanted to bring me here. We've MADE OUT before but now I'm the MAID OUT. I guess it's all in the spelling". She laughed and said "Okay let's get this mess cleaned up". She had a great sense of humor and any other time I would have appreciated it. It's much funnier now than it was at the time. She went down to the office and found an open store room. She came back with a broom and clean towels and linens. She said there was no vacuum cleaner so we would make do with with what we had. I doubt that room had ever had such a going over. Once we cleaned it up and remade the bed …. we did not mess it up. We parked back at her dorm and talked for a long time. Finally, I said "Okay I've got to ask, why were you willing to go with me to the motel"? She said "Because I know you and I trust you ….I knew if we started to go too far and I said NO you would respect me". I said "You were correct". I got her address, promised to write and we kissed goodbye. Before I drove off she leaned in my window, kissed me again and whispered in my ear …..

"I'm not sure I would have said NO".

GRADUATION DAY … FLY OVER AND ALL

I know it sounds corny but I really do get a lump in my throat when I hear a band play "Off we go into the wild blue yonder". The official name of that song is U.S. Air Force but most people know it for that opening line. General Hap Arnold was the person who thought the Air Force needed their own song. Actually General Arnold was a friend of my uncle John (when he worked at Bolling Field in Washington). When we finally graduated and were able to have those gold bars pinned on our shoulders, the band played that song and four jets streaked by low overhead. The feeling cannot be described….. you have to experience it. We all threw our hats in the air (just like the service academies). Some guys ended up with the wrong hat. I managed to keep my eye on mine and retrieved it when it landed. I had already arranged for the NCO who I spent so much time with on the drill tower to pin on my bars. In reality I had pinned them on myself, under the OTS boards we wore on our shoulders. That way I could pull off those boards and be one of the first to display the rank of Second Lieutenant. I still gave the NCO the dollar for luck and he went through the motion of pinning on the bars for a photo. Truth be known those bars had been on my uniform for three days before we had that final parade.

We discovered after we became commissioned officers that we had been discharged the day before as Staff Sergeants. They probably don't let you know that on purpose. You see, when you get your hands on that official "Discharge" paper it actually means that had you been able to have that record earlier you could have walked away and not owed another four years. For pay purposes all OTS students were given the rank of Staff Sergeant. Before we could be commissioned as officers, the Air Force had to first discharge us. The date of that official document was the day before we were commissioned. For a period of around 15 hours we had been Honorably Discharged civilians.

ROAD TO NEW ORLEANS

My parents had arrived late actually well after midnight. I'm not sure I told them that a few hours earlier their room had been a total wreck. I do recall getting back at the three guys who had trashed it. They got a bucket of water wake up call at 4 a.m. After I got the ring leader he helped with the other two. It did not matter ….. not many slept that night.

Immediately after graduation many of the newly crowned officers took some leave. We got 30 days leave (vacation) per year and you could draw ahead all 30 if you wanted. I would say most all decided to take at least two weeks. After all, we had been through a tough three months. We graduated on March 26, 1963. My orders read that I was to report no later than April 28 which meant (allowing for the 3 days travel time in the orders) I could take the full 30 days leave if I wanted. The NCO Drill Instructor that I had befriended had given me advice saying "When most of your classmates graduate they will be sick of what you all are going through and they take leave before going to their first assignment …. don't do that. There will be at least five or six who are

Administrative Officers just like you, all going to the same base. Get there first and you can probably have your pick of the jobs which are open. Trust me, some are good and some are bad. If there is a smart Personnel Officer, he or she will point you toward the best assignment. They always know". Best advice I ever got.

I told my parents what the NCO had said and that I really wanted to try and follow what he had suggested. My dad looked at my orders and said that it looked like I only had three days travel time. That did not give me time enough to drive them back to Birmingham so maybe they should just try to fly back. He said that we sure would not be able to spend the night in New Orleans like we had planned. I thought about it and finally said "No let's head for New Orleans, spend the night and go to Birmingham later on Sunday. I will leave for Amarillo either Monday or Tuesday depending on how tired I am. I have three days travel time and if I use a couple of days leave I bet I will still arrive at Amarillo before any of those other guys.

Graduation exercises started at 8:30 in the morning so we were finished by noon. I was already packed so we hit the road to New Orleans. Back then the drive from San Antonio took around ten hours so we did not arrive in the "Big Easy" until late at night. Things don't get started there early anyway, so even though we were all tired the adrenaline was flowing. We only paused long enough to check into a motel, shower, change clothes and we were ready to do the town. My dad had found a nightclub tour which sounded pretty good so he made reservations for the three of us. I had only been to New Orleans one other time when I was in college. Being under age at the time I had not seen much, other than to walk down Bourbon Street. In those days, as now, you could see quite a bit by just walking in the French Quarter.

The night club tour sounded like it might be neat. Now I am sure the first thing which comes to mind is how "neat" could it be for a 22 year old, single male to be celebrating in New Orleans night spots with his parents. Well first of all I was an only child who was fortunate enough to have parents who enjoyed a good time. For most of my growing up years they had allowed me to tag along on conventions and fun trips which they had gone on frequently. They loved to dance and party. Mom did not drink but my dad did. This time I could legally join him.

The tour bus picked us up at the motel. We hit everything from Pat O'Brian's to the old My-Oh-My Club (which was one of the first to feature female impersonators). We were having a good time. As part of the cost of the tour a watered down drink was included at each establishment. It was in the second club that I noticed a couple with our group who had brought their daughter along and she was a doll. By the time we got back on the bus I had found out they were from Canada. I had not noticed (she had) that they had joined the tour at our motel and were in fact, also staying there. For the rest of the nightclub tour we sat together and even danced a few times. She was 21 so we also had drinks. By the time the bus rolled back to the motel we were well acquainted. She was staying, as was I in the same room as her parents. I told my parents that the two of us had planned to meet in a few minutes to sit and "talk" in my car. I waited almost an hour and was about to give up when she showed up. She had decided not to discuss such a rendezvous with her parents and waited until she thought they were asleep. I have a fond memory of Canada and even till this day I have a pleasant thought when I hear their national anthem, "O Canada".

HEAD TO AMARILLO

We got into Birmingham and I had about enough time to talk to a few family and friends on the phone before I packed for the trip to my first assignment. When I say packed, I do mean "PACKED". I doubt I could have gotten another thing into that Comet convertible. The trunk was full and so was the back and front seats. I remember thinking that as far as living at home was concerned … it was now part of my past. I packed dumb stuff which I never needed and forgot things which eventually I had to replace. It was a scary, lonely feeling when I pulled out to head West. As far as I knew at that time, I might never be a resident of Alabama again.

The trip to Amarillo was rather uneventful except for a near disaster in a little town in Mississippi. This was early in '63 so the Civil Rights demonstrations were just beginning. A chapter in Mark Childress' book *"CRAZY IN ALABAMA"* reminds me a lot of what I ran into as I was driving through the middle of Forest, Mississippi. It may not have been Forest. It could have been Newton, or Morton or any one of the small towns along U.S. Highway 80. They all looked similar but for some reason, Forest rings a bell. A group of Negros were marching down the middle of the main business street. Traffic was trying to move through the marchers at a rather slow pace. All of a sudden the crowd began to dart in all directions. Rocks were being thrown both at the group and by some of the marchers (who must have anticipated trouble since they were armed with rocks of their own). If I had been asked, I could not have honestly said who it was from which group that "cast the first stone". Shouts of profanity were heard above screams and horns blaring. My car was suddenly surrounded by angry blacks pounding on the hood, the trunk and sides. My convertible top was down but no one made any attempt to reach inside or climb in, Maybe all the junk I had piled on the seats discouraged them. Remember my Comet was a four in the floor stick shift. I instinctively pushed in the clutch and stomped the accelerator. The roar of my engine sent those in front of my car diving to both sides. I saw a clear opening, dropped into first gear and left the smell of burning rubber for about 50 feet. There was another car stopped in front of me but I somehow managed to pass it on the left, barely missing a police car coming in the opposite direction. I did not slow down until I got to Jackson. At a filling station I found that other than a couple of small dents, possibly from rocks, I had escaped without major damage. I was not even aware of rocks hitting my car. All of the TV coverage of the LA riots of the early nineties and the Denny beatings brought back memories of just how lucky I was. Thinking back, I was scared but not half as much as I should have been.

The rest of the trip was kinda boring. I decided to swing North and head for Amarillo by way of Oklahoma City where I could spend the night pretty cheaply. One of the benefits we had heard about during our training was that we could stay on any military base at the Visiting Officer Quarters for next to nothing. It was my first of many experiences of taking advantage of that perk. I believe my room for the night was all of $2.00. It had been a long day so I collapsed and slept soundly.

I left early the next morning and rolled into Amarillo before 11 a.m.. Hard to believe but I had made it from San Antonio, to New Orleans, to Birmingham, to Oklahoma City, to Amarillo in my three days travel time. And back then there were no interstates. When you are young you can do that sort of thing. I decided to waste no time in reporting in. Not knowing what to expect, I was extremely nervous. After all I was a brand new Second Lieutenant and had heard many horror stories of what could happen to "Shave Tails" or "Brown Bars" as we were called (among other things), My orders were to report to the Base Personnel Office to receive my first assignment. I was hoping the advice about getting there first would pay off. If I got to pick from several jobs which were open maybe I could help determine my own destiny. It did pay off.

REPORT TO FIRST ASSIGNMENT

Base Personnel was not exactly fancy. It was a typical World War II barracks type building with so many coats of paint that various layers provided a ripple design on the woodwork. The thought crossed my mind that it was probably like the rings of a tree … if you could count the layers you could probably determine the age of the building. As you can see my mind was trying to do anything to divert my nervousness at reporting for my first assignment. It is fear of the unknown. This has to be the worst. Your stomach begins to knot up and you don't have a clue as to why. I said to myself, " What could be so bad about this"? Here you are in the middle of the most desolate area you've ever seen, about to get assigned to a job you haven't the slightest idea how to do, and you are nervous …. about what? The Assistant Personnel Officer, whose office I was told to report to, turned out to be a First Lieutenant. I walked in, snapped to attention, saluted (as we had been taught at OTS) and said "Second Lieutenant William Capps reporting as ordered Sir". Lieutenant Turner, the Assistant Personnel Officer burst out laughing and said "Who the hell told you to do that"? I stammered something like "We did it like that at Officer training School and you got demerits if you forgot". Turner said "Well Bill, it is Bill instead of William, right"? I nodded. "Bill, welcome to the real Air Force. You do report and salute to Senior Officers but rank among Lieutenants is like virtue among whores. My name is Frank Turner but you can call me C.T.". (Frank had red hair and I found out later the C.T. came from carrot top). He went on to say "You've not drawn the worst base in the Air Force but it damn sure ain't the best either. We say it is not the end of the world, but you can see it from here. Everything is so flat, if an airman goes AWOL you can see him for three days". C.T. told me there were six slots open for an Administrative Officer and since I was the first to report in I could choose any one that I wanted. All the newly commissioned Second Lieutenants who were supposed to report to Amarillo were shown on the same set of orders so he knew I was the first to show up. I told him that I did not have the slightest idea what an Administrative Officer even did as that was not something we learned at OTS. I told him that maybe he could offer some advice as to which job I should take. C.T. seemed to like my honesty and I think he liked the idea of telling me which decision to make. He said "Well it depends on how much you want to work and how involved you want to be". I told him that I wanted to learn as much about the Air Force as possible and I would just as soon stay busy. I only planned to be in for four years and wasn't planning on a career in service but I had been raised to believe you usually get more out of something if you put more in.

My remarks caught his attention, and he responded with "There is one position open in a permanent party squadron. The others are all student squadrons". I asked what the difference was. He explained that Amarillo Air Force Base was a Technical Training School and part of the Air Training Command (ATC). I would find out later that Amarillo also had a Strategic Air Command (SAC) Wing but they were isolated from the rest of the base and most of its staff had the attitude that

SAC was the REAL Air Force and ATC was just a necessary function. SAC flew "B-52's".

C.T. went on to explain that in Tech School there were basically three main permanent party squadrons: Headquarters Squadron and two Instructor Squadrons. All the rest (around 20 or so) were Student Squadrons. The members of these student squadrons would only be assigned to Amarillo for the period of time it took them to complete Tech School and that time varied depending on what school they were in. Each student squadron did have a few permanent party personnel consisting of the Squadron Commander, the Administrative Officer (formerly called the Adjutant), a First Sergeant and a couple of airmen for record keeping. The big advantage to starting out as Administrative Officer in a student squadron was the opportunity to move up to Squadron Commander. Most of the First Lieutenants who currently had those positions had been an Adjutant first. Many of them wanted to make a career in the Air Force and wanted to have Squadron Commander in their records (even if it was commanding a bunch of teenage students).

C.T. said to me that he felt if I worked in an Instructor Squadron I would come in contact with personnel who were all permanent party (with most being Career Officers, NCO's and Airmen) that I would be exposed to more of what the Air

Force was all about. That would be one of the reasons he suggested I choose the 3325th Instructor Squadron for my first job. The other reason he felt was more important.. He said, " As the Admin Officer for the 25th you will get to work for the guy I think is the best damn Major and one of the nicest guys on this base ..... Major Ralph Angus. You will love working for him".

Note: Ralph Angus is one of the real names I use because he is not a composite like many of the characters plus he was truly an influence on my years in the Air Force. I did indeed enjoy working for him. Turner was absolutely right.

I responded to Lieutenant Turner " Okay send me to the 25th Instructor Squadron". C.T. said "There will be plenty of time for that tomorrow, first let's get you set up in the Bachelor Officer Quarters (BOQ) and show you around the base". I didn't tell him, but I had already driven around the base just following signs so I would have an idea where the BOQ was, the Officer's Club, the Base Exchange, etc. I had sort of gotten somewhat familiar with the lay of the land before I reported to his office. His offer, however, was accepted and he rode with me in my car to check into the BOQ.

FIRST PERMANENT "HOME AWAY FROM HOME"

The young airman on duty in the assignment office was reading a copy of "*Playboy*" when we came in the front door. I guess I should say not reading but looking at pictures. He did not get up when we came in and I could tell by the look on C.T.'s face he was not pleased. He snapped at the young airman and said something that I did not hear. All I could think of was, oh great he is going to piss off this kid who probably has the power to put me in the worst room in the BOQ. The kid took two sets of my orders. You start out on new assignments with a huge stack of duplicate orders. You soon find out why. Every place you go someone grabs two or three copies. The Air Force could probably save millions of dollars (and many trees) by limiting duplicate files. My suggestion would be ,Read 'em, Act on 'em Pass 'em on …. then File 'em in one place. Sounds pretty simple … but NO, everyone wants their own set, plus an extra in case someone else wants a copy from them. CYA (cover your ass). A sergeant once told me if he did not see at least five copies of an order , signed by a Colonel, it probably wasn't that important.

Lucky for me the Officer's Club was next door to the BOQ and it was lunchtime. C.T. said "I'm going to run over to the club and see who all is there for lunch while you get your room assignment. I'll get us a table with some of the other junior officers. Just get your key, check out your room but then come on over to the club. You can meet some of the guys and after lunch we will come help you move in your stuff". The young airman had read my orders and asked "Do you happen to know which squadron you will be in"?. I told him the 25th Instructor Squadron and asked if he knew anything about it. He smiled and said "Only thing I know is my squadron whipped their ass in softball". Now he was talking about something I could relate to. I told him that I had played intramural softball in school and I was glad to hear that Amarillo had a league on base. He quickly warmed up to my interest when I asked him which position he played and how his team was doing. Before long he was talking about how he was going to try and get me a good room. Like everything in the military there is a pecking order. BOQ rooms are no exception. The largest rooms were assigned by rank and in some cases date of rank. There were only a few higher ranking officers living in the BOQ. They are called Bachelor Officer Quarters but they are not just for unmarried officers. Several of those living there were Captains and higher grade officers who were married, but for various reasons their families were elsewhere.

The quarters were basically in three different configurations. The largest had an outer living room, a kitchen, a bath and a bedroom. There were only about ten that size. They were taken up by a couple of Lieutenant Colonels, several Majors and a few Captains (who had been at Amarillo for a long time). The next size was in a "U" shape. When you walked in the door there was an area about five feet wide by twelve feet long. This had one couch on the right facing a wall across from it. When you walked around that wall and to the left there was a single (bunk type) bed. Across from the bed was a closet and on the same wall a door leading to a compact bath. I say compact because in a space of about four foot by five foot there was a commode, a sink and a shower stall. It was so tight you could sit on the commode, wash your hands in the sink and put your feet in the shower. In the corner outside the bath was a small refrigerator. The airman said normally this room is for a Captain and above but he thought he could assign it to me. I'm thinking if this is what is normally for a Captain and above I could only imagine what a brand new Second Lieutenant was supposed to get. Naturally I said "This looks great to me". He flipped me the key and said "Good Luck". The best thing about it was the fact it was on the first floor of the two story BOQ which only had stairs. I walked in, sat on the bed and bounced a couple of times. Not real comfortable but better than the bunk I had in San Antonio. I locked the door and headed over to the Officer's Club.

NO SUCH THING AS A FREE LUNCH

The Officer's Club at Amarillo had a personality all its own. It was large with several different rooms. One was a game room with several card tables and a pool table. Others were for private dining and meetings. The bar was separate from the large dining area. That area was full by the time I walked in. Finding C.T. was not easy. He finally saw me wandering around and motioned for me to come over to a round table where several Second and First Lieutenants and one Captain sat. He introduced me around. The only one whose name I caught right away was First Lieutenant Randy Garner. This was mainly because when he shook my hand he said "Welcome to Amarillo and Roll Tide". He was actually from a small town in Mississippi but he had graduated from the University of Alabama. C.T had told him I was an Auburn graduate. Randy and I hit it off right away (in spite of the intense rivalry between the two schools there is a tendency to join forces when thousands of miles from home). The Captain was a lawyer from the Judge Advocate General's office. His name was Morris Thomas. I would eventually have more dealing with him than any other person I met that first day. With the exception of Thomas and C.T. all the others at the table were from Student Squadrons. The conversation was centered on an upcoming parade which was to take place on Saturday. The general feeling was that it was going to screw up the weekend. I found out that most of the single guys tried to get the hell away from the base on the weekends. Sounded a lot like OTS and Lackland. The parades , even though less frequent than at Lackland were nevertheless competitive and dreaded by the participants. The Student Squadron guys at the table said that I did not have to worry as much as they did because I would be with permanent party people and there was not much pressure on them with marching and drilling.

No one seemed to be in a hurry so lunch went on for over an hour. Finally Captain Thomas said he had to get back to the office and everyone else chimed in that they had to go. One of the guys said "Well Capps you know it's the custom for the newcomer to pick up our checks". C.T. said "Don't let him bullshit you, 'bout the only way a Second Lieutenant gets stuck with a check is when we match for it". Three of the guys including the one who said I should pick up the checks all said they were in… and took out a coin. Not wanting to offend anyone on my first day I said "Okay just tell me what y'all are doing and I'll play". C.T. took out a quarter and said "We all flip a coin, if one ends up odd, he pays for everyone else".All five of us flipped. There were two tails and three heads. C.T. said "We keep flipping until there is only one odd". We flipped maybe two more times until the guy who originally said I should pay, lost. Everyone said "Serves you right" and handed him their checks. He reached over and picked up mine and said "You're going to be just like Garner. All you bastards from Alabama are lucky". I wasn't sure why he said that since Garner had not been flipping and I was pretty sure he had said he was from Mississippi. I found out later that the guy who lost was the Commander of the Student Squadron next to the one Garner was the Commander of and they were extremely competitive in everything. Also he had graduated from Oklahoma and they did not like the University of Alabama (they had just lost to them in the 1962 Orange Bowl).

So on my first day I had met some nice guys, had a new job assigned, a room to stay in, and had my first lunch paid for … so far Amarillo did not look too bad. C.T said "Come on. I'll help you move your stuff in and tomorrow morning I will take you to meet your new boss".

I'M MOVING IN AND THE SNAKES ARE MOVING OUT

As C.T. and I started moving my junk from the car, I noticed two guys wearing funny white outfits with hooded hats sorta like bee keepers wear. They were going under the crawl space beneath the BOQ building. They appeared to be carrying spray cans and long poles with rope loops on the ends. C.T. said "Looks like Rattle Snake Roundup" again”. Somehow I realized why that nice room on the “First” floor was vacant. We piled all my stuff inside. C.T. said “I will pick you up out front at 8 tomorrow morning”. As I made the last trip from my car I saw the two guys in the white coveralls come out from under the building with a huge rattlesnake. All I could think of was the holes I had seen next to pipes coming into my room from the floor. That first night I stuffed socks into them until I could find something better.

THE TWENTY-FIFTH INSTRUCTOR SQUADRON

C.T, and I actually walked to my first assignment. It was only about four blocks and he said parking spaces were hard to come by. There were four barracks buildings set in a square with a grassy quadrangle in the middle. The buildings all looked alike. In fact most of the buildings on the base looked alike. All entrances were on the inside facing the open area. A paved walkway went around the quadrangle. It was large enough to be a driveway but a sign where it connected to the street said “NO VEHICLES”. The four buildings were conveniently located near the center of the base or at least toward the center of where most of the courses were taught. It was also close to the office of the Commandant Of Troops. The two story barracks buildings were split in the center by what were called Day Rooms. Airmen and NCO’s occupied rooms to the right and left of these. The Day Room on the first floor was set up as a lounge, with a TV, card table, couches, chairs, magazine rack and a ping pong table, On the second floor where another Day Room used to be had been divided into offices. There was a long office across the front and three small offices across the back. You entered the smaller offices through one door in the back wall of the larger office. When you entered this door you walked right up to the desk of the Squadron First Sergeant. To the young airmen this was the most feared man in the unit. He kept the duty rosters. He assigned all the good stuff and bad stuff and he could also strongly influence punishment if you got into trouble. He sat behind an over sized desk facing that middle door which was always open … except when an airman was in trouble in which case it usually closed with a loud slam. When C.T. and I entered the large office, three airmen who were all seated behind desks all jumped to their feet. One of them sang out rather loudly "ROOM, TENGE HUT"! That was the phonetic way the word "ATTENTION" had been taught at Lackland. I have to admit I felt a bit embarrassed. This was the first time a room had been called to attention for me. Well it was really for C.T. who was leading the way. He casually said "As you were" and they all sat back down. He asked an Airman First Class who seemed to be in charge if the Major was in. He replied that he did not think so but that

Sergeant Summers was. By now Sergeant Summers had appeared in the doorway of his office. He said "Welcome to the 25th … you must be the new Adjutant". C.T. laughed and said "You'll have to excuse Sergeant Summers. He's left over from the old Army Air Corps. He has not accepted Administrative Officer as a title yet".

We shook hands and he said he would show me my office and help me get acquainted with the staff. He explained that Major Angus was not in but would return in the morning. C.T. said he was leaving me in good hands and was heading back to his office, if I needed anything, just give him a call, Sergeant Summers introduced me to an Airman First Class Boyers who was in charge of the administrative staff. His assistant was an Airman Second Class by the name of Roberto Quandro Dias Rolando. For my first meeting, however he was introduced as Robert. I did not find out all of his given names until several days later when he told me he was from Cuba and that the Air Force could not put all his names on his ID Card. The third Airman on the staff looked pretty old to be an airman Second Class. He was small and skinny and had a rather drawn looking red face. His name was Charlie Briggs. I would find out later that he was the epitome of this new dysfunctional family I was becoming a part of.
Sergeant Summers took me into my office. I liked the sound of "My Office". I felt like it was a big

deal. Here I was, a brand new Second Lieutenant in the United States Air Force and I had my first private office …. complete with a window … and a view. That is if you can call "a view" looking down on the walkway which airmen used to march to class each day. At any rate it was mine and it was fairly large, the same size in fact as that of the Major's, which the sergeant let me walk into. He said "Major Angus told me to tell you to leave early. He said he knew you need to get settled in and unpack and that he would see you in the morning around 8 o'clock. That sounded good to me so I stuck around a few more minutes said goodbye to everyone I had met and headed back to my room. I actually went back and got my car and drove to the Base Exchange. I made my first purchase which was a portable TV (black and white of course). With rabbit ears I could pick up three channels. I already knew with my next pay check I would buy a stereo …. I had brought all my records from home (mostly 45's).

FIRST DAY ON THE JOB

I woke up early, I guess with nervous anticipation. When I say I woke up early I mean real early … around 3 a.m. I finally got out of bed around 4, showered, shaved and put on my best pressed uniform. Sergeant Summers had said be at the office around 8 a.m. so to kill some time I decided to drive up near my office. If I saw a place to park then I would drive instead of walking the four or so blocks. I didn't want to arrive on my first day all sweaty. There was a parking lot close to the barracks. When I drove in I noticed only a few which were apparently reserved places. One immediately caught my eye. It read COMMANDER 25th INSTRUCTOR SQUADRON. Next to it was one for the Commander of the 24th Instructor Squadron. There were several more which read Instructor Of The Month with various schools such as Administrative School, Missile School, Fuels School, Air Craft Maintenance, etc. I looked hard for one which said Administrative Officer or Adjutant but there was not one. I happened to see an Air Police Truck pulling into the lot so I rolled down my window and asked the Airman where I should park since I was coming to my new assignment. They both laughed and the driver said "Anyplace which ain't marked reserved for somebody … especially one which says Commander". It was still early so I went back to the Officer's Club to have breakfast. My stomach ached more from nerves than my breakfast but afterward I headed back to the parking lot and found an empty spot toward the back of the lot. As I walked into the orderly room there was only Airman Briggs sitting at his desk. He did not exactly hop to attention as I came in. As a matter of fact he just sort of looked up grinned and said " How's it hanging Lieutenant"? Somewhat taken back I stammered something like

“Not bad, how are you this morning”? Charlie did not answer and just went back to whatever he was doing. I walked into my new office and sat at my desk, thinking okay you’re here might as well make the best of it, knowing full well I did not have the slightest idea what the hell I was doing.

Sergeant Summers arrived within minutes and came walking into my office. He said” Coffee will be ready in about five minutes, how you like yours”? I told him I was not a coffee drinker and he just mumbled something about you can’t survive around here without coffee. I found out later that he alone consumed about 20 cups a day. I doubt I ever saw him without a cup in his hand. I sat there wondering how in the world do I start when I am completely lost. My thoughts were interrupted when in walked Major Ralph Angus. I hopped up and immediately popped to attention and I probably saluted as a matter of habit from OTS. I said “Good morning Major’. Before I had a chance to introduce myself he chuckled, stuck out his hand to shake and said” None of that is necessary around here. As long as we are in the office or at my house you call me Ralph. You only call me Major when we are out in public, otherwise we are pretty informal”. What Major Angus said next stuck with me all these years. It came when I said that OTS did not teach me anything about being an Administrative Officer and I noticed a bookcase in my office filled with regulation books … should I start studying them? He said “Those regulations were written for people who couldn’t think for themselves, use good common sense and only read ‘em when you need ‘em”. I knew right then I was going to enjoy working for him. I was fortunate being exposed to someone like Major Angus for my first job in the Air Force. He probably had a bigger influence in my ability to work with people than any other person. He was a quite, modest person who was well liked by everyone on the base, I say modest because Sergeant Summers told me later that the Major was probably the most highly decorated officer on base but he did not wear all the ribbons and decorations he was allowed to put on his uniform. He was a Command Pilot who had flown with the RAF before being commissioned into the Army Air Corps (before it became the U.S. Air Force. He actually had ribbons and medals from two countries which he was authorized to wear. I always thought he refrained from wearing them so he would not look more important than some of the other senior higher ranking officers on base. He was shy about talking about his accomplishments.

After three days of talking to Major Angus, I had already formed a significant opinion. The best decision I had made was following the advice I had been given and getting to Amarillo early. It was because of that I got assigned to the 25th and had Major Angus as my first boss. My only real concern was with all the time he had in service, why was he still a Major? I kept meeting higher ranking officers who were not nearly as sharp. This would end up to be an important influence on deciding if I wanted to make the Air Force a career. At that particular time, however, I was only thinking of spending my four year commitment and getting back to civilian life. Two years down the road I did give serious thought to staying twenty years.

I got quickly involved with every facet of my new job. My plan was to stay as busy as possible so the time would pass faster. When you volunteer for everything the Air force will give you all you can handle and more. The Major told me that officially I was second in command of the squadron. I would be assigned many “Extra Duties” mainly because someone had to be responsible and it was normal to throw most all of the crappy stuff to a Second Lieutenant. Since he had been without an Admin Officer he had assumed the extra duties himself but he was now most happy to pass those on to me.

My first set of orders for “EXTRA DUTIES” were a shocker. It would only prove to be a fraction of those I would be assigned over the next twelve to fourteen months.

When I first looked at the order I thought how in the world can one person be expected to do all those different things especially not knowing what any of them meant let alone how to do them. Sergeant Summers said "Don't worry I will walk you through what you have to know, which ones to worry about and if I don't know they must not be too damn important so just bluff your way on those. A lot of times someone comes up with a title to justify their job and they make each squadron send them a report so they can file it in a desk drawer. They don't mean squat". Summers was definitely "Old School" and he had a way with words. Here is a list of those first "Extra Duties" I got that first week:

Fire Marshall, Ground Safety Officer, Conservation Officer, Voting Officer, Information Services Officer, Personal Affairs Officer, Bond Officer, Athletic Officer, Unit Postal Officer, Records Officer, OJT Officer, Collateral Training Officer, Special Services Officer, Insurance Officer, Historical Officer, Unit Security Officer and a Member of the Squadron Promotion Board.

Many of those you see required little of my time with maybe an inspection here and there or cutting an order once a week or attending a monthly meeting. There were some which required keeping strict records or charts on our bulletin board (more about those later). Then there were others such as my favorite, Athletic Officer which not only took up a lot of time but allowed me to get involved and play sports.

WE ALL LOVE A PARADE … HA HA

The guys at my first lunch were right. That first weekend there was a parade I got the chance to lead the squadron and perform an inspection of the troops. Apparently when someone retires they can request a Retirement Parade. The retiree or in the case of that first parade, several retirees get to sit in the reviewing stand with the Base Commander and the Two Star General who was the Tech Training Center Commander. As it turned out one of the retirees was an NCO from the 25th Instructor Squadron. He was well liked so at least the guys who were to march in our Flight had somewhat of an interest in being there. Most of the time these Reviews or Parades were hated by our squadron members. Let me explain. The 25th Instructor Squadron was one of (if not the) largest permanent party squadrons on base. The large majority were career high ranking NCO's who were experts in their field and were the instructors in the various schools. They taught everything from Administration to Refueling to Missile Electronics. We even had 44 officers who headed up those schools, ranging in rank from First Lieutenant to Lieutenant Colonel (all of whom outranked me). The last thing any of these guys wanted to do was march in a parade. The squadron was so large we had five flights in each parade. The large majority of our squadron members were married and lived both off base and in the base housing section. We did have single airmen and NCO's living in the barracks where our office was located and also one of the other barracks buildings in our quadrangle. The personnel from both of the barracks buildings usually got stuck with being in the parades. Since we did not have enough men to fill up all the Flights from these two buildings, the balance came from Sergeant Summers famous "Duty Roster". When you got on "His List" it was hard to get off. With as many men as we had there was always a sufficient number who had committed some sort of misdeed to be included on the sergeant's list. Having to march, especially if you were permanent party was perceived as a punishment.

I did know how to lead the troops but being thrown in on my first weekend as the brand new member of the squadron for something they all hated, was not something I looked forward to. We had a practice Friday afternoon and surprisingly it went well. One of the Master Sergeants who had gotten on Winter's Shit list for something that happened at the NCO Club came to me and said "Lieutenant if you will let me I will get these guys in shape and you won't have to worry about inspecting them. All you will have to do tomorrow is lead them when we march past the reviewing stand". I thanked him and said "That sounds like a great idea". I remembered what my Drill Sergeant buddy had said "Get on the good side of your NCO's and they will take care of you". He took charge, our practice went well and Saturday went off without a hitch. We would not by any stretch been named honor flight had we been in competition, especially when we passed the reviewing stand and a bunch of the airmen who had worked for the NCO who was retiring whistled and called out his name. Monday morning Major Angus came in my office and said I heard you handled your first parade well. Sergeant Summers gave me a report. All the men liked the way you took command. I learned early on, just say thanks and don't give details if not asked.

As a Commander the Major made it clear that he had an "Open Door Policy". If you had a problem and wanted to see the "Boss" he was always available. It was also clear that "Door" was only accessible by passing through the office of the SERGEANT MAJOR. Master Sergeant Joseph E. Summers. He also made it clear (without Major Angus knowing) that it better damn well be something he himself could not handle before you bothered the Major. The "pecking order" worked that way too …. Sergeant Major, Administrative Officer then the Commanding Officer.

MY FAVORITE EXTRA DUTY

Long ago in this book I said I did not have real athletic talent. However I love sports. I was not great in any sport but I think I was a little above average. In college I played intramural sports ranging from flag football, fast pitch softball, track and even pitching horseshoes. I bring this up as background so you will understand that at this time in my life in Amarillo, Texas I was about to discover athletic talents which apparently had been dormant in my body waiting to be released. Keep in mind that I had just finished three months of intense training at Officer Training School. I was in the best physical shape I had ever been (and have never been there again). MY first week at the 25th I found out the importance sports played on the base. The biggest deal around was something called the "COMMANDER'S SPORTS TROPHY". This was awarded by the Two Star General who was Commander of the Tech Training Center. Every squadron competed in a variety of sports, gaining points on how you finished in each. The squadron with the most points at the end of the year won the coveted trophy. Squadrons received some points just by fielding a team; therefore it was important, NO it was imperative that you have a team for every sport, regardless of how good they were. It was to this end that I received my first real order from Major Angus. "As our Squadron Athletic Officer, you are responsible to make sure we have a team representing the 25th in every sport. I don't care if we win or not but I want the minimum points for fielding a team". Even though various sports covered a period of over a year and a half I am going to try and explain many of them in the following pages. I accepted the order from the Major with enthusiasm, even treating it as a challenge. This was something that I could have fun doing. Sure …. but no one told me the politics of how big a deal this really was. What I found out real quick was mind boggling. If you had some good athletes in a particular sport (and with one of the biggest squadrons we had quite a few) they would be transferred for "temporary duty" to the Headquarters Squadron during the season of the sport in which they excelled. My first introduction to this was when I posted the sign up sheet for softball. When I started asking around as to who we had who would be a good fast pitch pitcher the same name kept popping up. We had a Staff Sergeant who was a windmill type pitcher and had been and still was a personal friend of none other than the most famous softball hurler of all time Eddie Feigner (King and His Court). I could not wait to talk to Sergeant Phillips or "Fireball Phil" as I was told he was appropriately nicknamed. When I called him to ask why he had not already signed up, he laughed and said "Lieutenant, didn't anyone tell you that they always transfer me to Headquarters Squadron during softball season".

This ticked me off but on the advice of the Major (who said it was his boss at Headquarters Squadron who transferred Phillips) that I should not fight City Hall. I decided to just set up the first practice and see who showed up. Sergeant Summers suggested that I appoint a coach from one of the NCO's living in our building. He said we had a Sergeant Flanders who in his day, had been a pretty good baseball player. He had even tried out for the old Minor League Memphis Chicks and had actually played a few games before the Air Force lured him away. I called Sergeant Flanders and he said he would love to be the coach.

Even though we had 44 officers in the squadron, I was the only one to show up at the first practice. We had 36 airmen and NCO's sign up but only 22 of those made it to the first practice. Coach Flanders took charge. He said, "Out here I am the boss … even our young Lieutenant will take orders from me, that is if he makes the team". This got a good laugh and broke the ice, It was obvious our coach was taking his assignment seriously. He said he really wanted to put together a winning team. Since all of the student squadrons teams had young airmen (in shape straight out of Basic), our task would not be easy. The biggest cheer came when Coach Flanders said "I don't really give a damn about beating anybody except Headquarters Squadron". He said that they had not only transferred Sergeant Phillips but also they had given three airmen who were all great athletes temporary assignments as well. " This ain't fair but we're gonna whip their asses".

Right off the bat Flanders had us running laps and wind sprints and I think we finally got around to throwing and hitting a few balls. At any rate by day three of practice we only had 11 people. He said we should each try to recruit at least one good ball player. He added "I have found us a good pitcher". We had in the squadron an older Master Sergeant who in his day had been on the ATC fast pitch Championship team. He had retired from pitching because he thought he was getting too old to keep up with the young players. Flanders said "He can no longer run but he can still fire a fast pitch. I have talked him out of retirement and Headquarters Squadron doesn't know anything about him".

His name was Sergeant Mack and we met him on our 4th day of practice. It was obvious he had been an athlete. He was only 50 years old (which to me at 23 seemed almost over the hill) but he was in great shape. He had no beer belly (like a few of his new teammates) and his arms looked like he still might be pumping a little iron. He said it would take him a couple of days to get his arm in shape. When we saw the first pitch we were impressed. All of a sudden we felt we might have a chance.

Most fast pitch softball teams are as good as their pitcher. If you have a good one you can be competitive. We left practice eager to recruit more players.

I had my chance the next morning. Sergeant Summers said that we had a new Second Lieutenant reporting into our squadron. He was coming straight from Officer Training School but because of his teaching skills in typing, he was being assigned as an instructor in the Administrative School. Major Angus thought it would be a good idea for me to help get him settled and show him around. Apparently what C.T did for me was not his normal practice but it gave me a chance to pay it forward. The first time I met Mike Rex, I liked him. Over the next year and a half we would become good friends. I am not sure how, but the subject of softball came up early and Mike expressed an interest to play. I believe as soon as we got him moved into the BOQ we went to the Base Exchange where he bought a glove. He went straight from there with me to practice. It felt good not being the only officer on the team.

At that practice we ended up with 14 guys. From that group Coach Flanders built a team which would eventually win our division but we would get knocked out in the first round of the base championship. During the season we played Headquarters Squadron twice and we split by winning the second one. We came back in the bottom of the 9th inning to tie and scored the winning run in the bottom of the 12th. Headquarters would go on to win the Base Championship. The transferring of personnel would not become a big deal again until Flag Football season.

OTHER SPORTS BEFORE FLAG FOOTBALL

I will keep football until last because it's the most interesting story but other sports took up some of my time because no matter what it was I had the responsibility of fielding a team so our squadron at least got the points for participation. There were a couple which yielded only a few points since they were considered individual sports. They were things like Archery where one person could represent the squadron, We had a Staff Sergeant Sullivan who actually came to me and said he wanted to compete in the Base Archery competition. Turned out that he was good at it and in fact it was the only Base Championship which the 25th could claim they won. He would later on volunteer to coach our football team. He also had a really cute sister who Major Angus told me he did not think it would be a good idea for me to try and date. I did not agree with his thinking but did comply with his suggestion.

Another individual sport (but one where two individuals from our squadron could compete) was horseshoes. I put up the sign up sheet in hopes we would get several wanting to compete. My thinking was we would have squadron competition and then our best two could represent us in the Base Competition. I was our only signee. I did talk Sergeant Summers into going with me and being our number two. I actually thought I was good at this sport (if you can call it that). My dad had always won the trophy for pitching horseshoes at his Masonic Lodge picnic. He was good. I still remember the day when I finally beat my dad, pitching horseshoes. For that reason I thought I was also good. When we got to the Base Competition they had real clay pits set up. Of course neither Sergeant Summers or I had been practicing. You drew names from a hat to see who pitched against who and in what order. My name came out first and I was to pitch against a Senior Master Sergeant. I was in civilian clothes he had arrived in uniform.

I pitched first got a ringer and a leaner. There was a surprising number of spectators so I must admit I was feeling pretty smug at my start. The Sergeant smiled and congratulated me then tossed his first shoe, a ringer knocking away my leaner and topping my ringer. The crowd cheered. They did again when he threw his second ringer. From then on my memory just fades. We played the best two out of three matches. I barely won the second. The third was not even close. He was getting warmed up and I think threw double ringers each time it was his turn. Of course he went on to win the Championship. I found out later this was the third year he had been Base Champion. After the meet was over, he invited me to the NCO Club and bought me a drink. He said "You do know it was just the luck of the draw. You pitched better than any of the other competitors. If you had not drawn me in the first match you probably could have gotten Second or Third". I had been the only one to beat him at least once.

When Tennis came along it was the same story. I could not find anyone willing to represent the 25th. With as large a squadron as we had I thought there would be plenty of tennis players. I am sure there were, they just did not want to compete. In the Orderly room it became a standing joke. Each morning for about a week the question was asked "Anyone sign up for tennis". On Friday the staff made a big deal of awarding me a racket (actually borrowed) saying" Lieutenant Capps won the squadron tennis tournament and will represent us ". I think the Major put them up to it as he was determined we would at least get our participation points. I had never played tennis. Where I grew up we did not have a tennis court. I only knew that in keeping score the term "Love" was used, so how bad could it be. I was soon to find out. As in horseshoes I had the "Luck of the Draw". I played the first match against a First Lieutenant who graduated from the University of Florida. All the single Lieutenants on base knew him from the Officers Club because he and his wife ate there almost every night and she was one of (if not the) best looking woman on base. They were both tan, blond and obviously from Florida. We all called them Barbie and Ken. I played Ken. Trouble was obvious when he said "Let's hit a couple to warm up". His first serve hit me on the shoulder before I could duck. I think we were to play a set of 6. Ken won 4 or 5 with me not returning a ball. Even on my turn to serve it either came back so fast I never saw it or I had a double fault. The final game he played me left handed and I still lost. I did stay for the whole tournament because I sat with his wife in the stands (hoping one of my friends would see me). She was nice and told me not to feel bad because her husband had won the Southeastern Conference singles title when they were both at Florida. Later that week, after he heard I graduated from Auburn he invited me to their house for a drink. He had a shelf full of tennis trophies. He had added another by winning the Base Championship. The only good thing was he was not in Headquarters Squadron.

Basketball season we had a large group sign up to play but no one wanted to coach. All of our tall players had been put on "Temporary Duty" in

Headquarters Squadron but we still had a few good players. I became the coach ….. in name only because I knew nothing about basketball. Growing up I did shoot baskets in my backyard (and yes it was a peach basket nailed to the barn wall) but being short in high school I was just a spectator. That is also what I was as the coach. Our team just played and had fun. We won our share of games but as usual Headquarters Squadron won the Base Championship. On our last game of the season we were way behind and there was about 30 seconds on the clock when the team started saying put the Lieutenant in. “You gotta go in and take the last shot”. So I went in ….and took the last shot. Loved to say I sunk one from downtown but NO, …. it did, however, hit the rim.

FOOTBALL … LONG STORY BUT A GOOD ONE

Sergeant Sullivan had asked if he could coach our football team. He said “Lieutenant, I know about Headquarters transferring many of our best players but we still have a bunch of guys who can play. I’ve already found us a quarterback. His name is Doss. He is an Airman First Class who lives in your building”. His name was Brian Doss. He was a young black kid from Oklahoma. The fact that he lived in our building was unusual. The airmen and NCO’s who lived in our two buildings were allowed to select which one they wanted to live in. All of the blacks except Doss had segregated themselves and lived in one side of our other building. This would create a problem which I will cover in a later chapter.

The football team shaped up fast. We had a fair sized offensive line which gave great protection to our quarterback. We had enough people to field an offensive and a defensive team. Remember back in 1963 even college players often played both ways. We also had a couple of the bigger guys who played on the line for both. The rules allowed plenty of substitution. As with most of our sports activities I was the lone officer from our squadron who showed up to play. Mike came to watch but decided not to play football. We had a surprising number of NCO's. This was because Coach Sullivan, who was well liked among his peers did a good job of recruiting them. He let everyone write down the position they would like to play. After a few days of practice he moved people around and tried them at various slots. After watching Airman Doss the first day, no one else even asked to play quarterback. I signed up to play end on offense and deep safety on defense. The latter I never did get to play.

I had to prove my ability to hold on to some of our quarterback's bullet passes. He said that I had pretty good hands and I secured one end position while a young black named Dee Anderson played the other. He quickly gained the nickname "SPEEDY DEE" Anderson and he was fast. I was in hopes that I gained my position because of my ability and not just because I was an officer. If I had been competing against Anderson, I would not have made the team. Since he and I spent a lot of time together working on plays, I got to know him quite well.

He was from Mississippi originally but his family moved to Michigan when he was in high school. I was shocked to find out he had a college degree. He was an Airman First Class and had moved up fast as an instructor in the Administrative School. I asked why he had not applied to come in the Air Force through Officer Training School as I had done. He said he had graduated from a small all black college and he doubted the Air Force would have accepted him into the program. He said "My Air Force Recruiter never mentioned anything about OTS. I actually hurried up and joined because many of my friends were getting drafted and I didn't want any part of the Army". I told him the only requirement which I knew about for OTS was he just had to be a college graduate.

After the season was over I talked him into letting me check with the Commandant of Troops about recommending him for Officer Training School. One of the most rewarding things which happened to me was the day Second Lieutenant Dee Anderson reported back to the 25th Instructor Squadron. He was assigned to take over the command of the division which a little over three months before, he had been in as an Airman First Class. He walked back into my office, saluted and said "Sir, I just wanted to thank you for helping me get to be an officer". I told him he could drop that "SIR" stuff. I will never forget his comeback He said "I'm sorry Sir … it's just force of habit". Meanwhile back to football. I got a little off track but Anderson becoming a Second Lieutenant is a neat success story.

I must mention that all the sports I talk about (practices, games matches, etc.) were held after working hours. As the Squadron Athletics Officer I did sign ups, recruiting coaches and equipment requisition during the day as part of my job. The football team practiced every afternoon as soon as everyone got off. Most everyone could get away by 4:45 or 5:00. The practice field was almost two miles from the office. I used to put the top down on my convertible and drive to practice. I would take as many guys who could pile in and on the back of the seat (usually 4 or 5). I had not paid much attention, but one afternoon I realized that I was the only white person on the offensive team. It became apparent when one of the Staff Sergeants, who played blocking back said "Only the offense gets to ride with the Lieutenant".

This was all taking place in the fall of 1963. Around that time racial tension was beginning to gain a lot of publicity. The Sergeant was about to take his place in the front seat (he often pulled rank on the others riding as they were all Airmen). He saw a copy of The Birmingham News (hometown paper which my parents had sent) lying on the front seat. He said, “I’m not gonna ride in a car and sit on a paper like this one”. It was obvious that he said it in a joking manner so I responded by saying “That’s too bad we’ll see you at the practice field”. Much to the glee and cheers of the airmen in the back, I drove off and let him walk. He managed to get another ride and arrived almost at the same time we did but he had fun telling everyone “The Lieutenant‘ s a pretty good guy but don’t say anything about his hometown or he gets upset”. That same Sergeant would prove to be my biggest ally when later on we had a few racial problems in our squadron.

As the season progressed we got better and better. We won game after game beating many of the student squadrons which were all filled with young “in shape” athletes. We just had smarter plays and the coaching was superb. It also did not hurt that our quarterback, when he got out of the Air Force managed to get a scholarship to play at Oklahoma State. Speaking of college I might point out that Headquarters Squadron managed to transfer (temporary duty of course) two Second Lieutenants to their team who had played football (yeah the tackle kind) at Texas. They had promised not to block too hard or at least to take off their Cotton Bowl rings during our "Touch" football games. It actually was not touch but flag football where each player wears a cloth strip from the waist on each side and one in the back I think it was held on with Velcro. The player is down when one strip is removed. Headquarters had done its usual job of loading a team.

HATCHET MAN HOBART

Behind every kind, gentle, soft spoken Colonel, who eventually becomes a General, there is most likely a Lieutenant Colonel or Major who rides up with him and handles the dirty work. I call it the Air Force version of "GOOD COP …. BAD COP". In the Tech School at Amarillo, we had that situation. The Tech School Commander was a well liked Full Colonel by the name of Dudley E. Favor. Colonel Favor was the idol of most of the young officers on base (at least those who hoped to make the Air Force their career). He was by all outward appearance and actions, Mr. Nice guy, and with all my dealing with him he truly was. He was a Command Pilot with a chest full of ribbons and the epitome of what all my Lieutenant peers wanted to be. He had lots of authority but did not seem to flaunt the fact that he had probably the second most important job on the base. The only one being greater was the Major General's job of Commander of the Tech Training Center. Colonel Favor did not personally handle any of the jobs which tended to get people mad. He had an "Assistant" for that. Actually Lieutenant Colonel Hobart's title was something like Operations Officer but the junior officers all called him "The Assassin Assistant", with emphasis on the "Ass". He had other nicknames like "No Heart Hobart" and "Black Bart Hobart", but you get the idea. My Lieutenant friends who were Student Squadron Commanders all had stories to tell about how much they hated him. I had seen him in action because I was now attending Colonel favor's weekly staff meetings. My boss Major Angus had been promoted to Lieutenant Colonel and was now working as an Administrative Assistant to Colonel Favor. Until a new Major could be assigned to the 25th Instructor Squadron, I had been named acting Squadron Commander. There was a bit of a nervous stomach for the first one I attended but with my old boss also there I quickly felt at ease.

It was at one of those weekly staff meetings that Lieutenant Colonel Hobart first brought up the Commander's Sports Trophy. He was giving a report about the Headquarters Squadron standings as far as the points were concerned. You see, Colonel Favor and everyone in that staff meeting accept the Major who was Commander of the other

Instructor Squadron and I were in Headquarters Squadron. Hobart said they had won this and they had won that. He even had a chart which tallied up point totals for participation, First, Second and Third in each event. I don't remember anything where they finished lower than Third in anything. Anyway he rambled on until Colonel Favor interrupted and asked "How are we going to do in football"?, which was the sport coming up. Hobart had the audacity to talk about how they had "Loaded Up" the team by transferring several top athletes from both of our Instructor Squadrons. He said "We are a cinch to go undefeated. We have the best team on the base". Lieutenant Colonel Angus saw the look on my face and he knew I could not keep my mouth shut. I spoke up and said "Colonel, Headquarters hasn't had to play the 25th yet and we might just give them some competition". Everyone laughed with the exception of Hobart, who just gave me an angry look.

Early the next morning Sergeant Summers met me at my office door and said "You need to get your butt over to Lieutenant Colonel Hobart's office pronto. He's already called twice to see if you were in. He sounded pissed about something". Hobart's office was about the size of Colonel Favor's, although he had a few more filing cabinets and there was a clutter of paperwork on his desk and a side table. He had a guest chair directly in front of his desk, which was no doubt purposely lower than the one he sat in. In fact when he asked me to sit down it felt like I was automatically sinking even lower so that I was forced to look up directly into his beady eyes.

He got straight to the point. He said "Lieutenant, do you know that if Headquarters Squadron wins the Base Football Championship that we win the Commander's Sports Trophy"? I said "No Sir, I knew from your report in staff meeting that Headquarters was doing well but I did not know that football could be the deciding points". There was a long pause and finally he said, " Well I'm sure you see the significance in the game coming up with your squadron …. you guys don't have enough points to place in the top five". I waited for what seemed like an eternity. I don't know how a cold stare can make you sweat but I felt like I was about ankle deep in perspiration. Finally I mustered up the courage to blurt out "Colonel are you asking me to throw a game"? He mumbled something like "Of course not, I'm just pointing out the consequences of what would happen if Headquarters lost to the 25th". I said "I'm sorry Colonel but it sounds to me like you want us to try and lose and I'm afraid that's not going to happen. We are going to try our damnedest to beat the hell out of Headquarters". There was another long pause. I had hoped a hole would open up and I could disappear but I had gone too far to back down now. I said "Is there anything else Colonel"?

He simply smiled that "I'LL GET YOUR ASS LATER SMILE" and said "No Lieutenant, that will be all .... for today".

I had never planned on a career in the Air Force and I am sure if I had, my encounter would have taken care of that. My office was full when I got back. Sergeant Summers had gotten a call from a WAF who worked just outside Hobart's office. She had told Summers that Hobart was raising his voice and I was talking back to him. I don't remember any of that happening. In fact, as far as I know, the way I told it above is exactly what transpired. Legends, however are built on far less and by the time I arrived the whole orderly room was buzzing. Remember, I was at the time the acting Squadron Commander. Sergeant Summers had called Sergeant Sullivan and he had rounded up a couple of other NCO's from the team The story as relayed by the WAF outside Hobart's office had been embellished as it was retold to each new arrival.

By the time I walked in, it was "FACT" that Lieutenant Colonel Hobart had told me in no uncertain terms .... No, I think they said he "Ordered" me to make sure the 25th threw the game against Headquarters Squadron. Of course everyone was excited because I had stood my ground and said I would refuse to obey such a stupid order. I tried to explain that it really did not happen that way but the more I thought about it I figured, why not let the rumor mill take care of itself. What better way to get our team fired up to beat Headquarters Squadron. Why even Alabama Coach "Bear" Bryant would have been proud. He was a master of the old bulletin board psychology.

LETTER TO THE COLONEL

It was not long until I began to worry about how things had transpired. The WAF had overheard just enough of what went on in Hobart's office to back up the story in case he denied the meeting took place. I felt some action was needed just in case we actually did upset Headquarters. Hobart was so irrational, I would not know what to expect as the fallout. That afternoon I wrote what would become one of the best letters I wrote while in service to Colonel Favor. In it I spelled out exactly what had happened. I said word for word (to the best of my recollection) what Hobart had said. I was specific that he denied requesting that we throw the game but I told Colonel Favor that regardless of the consequences, I could not compromise the way I was raised. There was a principle involved. I could only come up with one meaning for Hobart calling me into his office. Also I added if I had judged him (meaning Colonel Favor) correctly, I could not bring myself to believe he would approve of Hobart's actions. I almost got up enough nerve to send a copy to Hobart but thought maybe you can push your luck too far.

THE GAME

Kickoff was set for Saturday at 2:00 pm. We probably had the biggest turnout for any game all year. Practically every married member of each team brought their wives and kids. Friends, girl friends, other squadron members, the sidelines were filled. Tension was high, Lieutenant Colonel Hobart was there (in uniform). He appeared to be glaring straight at me. We were all (including Headquarters Squadron team members) shocked to see Colonel Favor also on the sidelines. He rarely attended sporting events. Everyone accept Hobart had on civilian clothes. He really stood out because he had on dress blues. Power and authority does funny things to some people.

The game progressed in a boring fashion. Each team would run three plays and kick. There were a few first downs by both teams but always the defense would rise up and make the big play. They got down close one time but we intercepted and ran the ball back to mid-field. The best defensive back that Headquarters Squadron had was playing against Dee Anderson man for man. He was a lot bigger than Dee and even though it was illegal to throw low blocks he would hit Dee as he came off the line. He would either knock him down or at least mess up his route. The guy played extremely rough and when he could get away with it pretty dirty. He tripped Dee twice but never got caught by the referees. On one particular route Dee and I both went long, he went up to hit Dee and accidentally tipped the ball back to me. I made a diving catch on about the ten yard line. Dee was knocked flat and as they both went to the ground some words were exchanged and a couple of punches thrown. Referees quickly broke them up. It was then that Dee realized he had sprained his ankle. It was also about the same time that another referee who was not even where he could see the play ruled my catch incomplete. He said I trapped it when I hit the ground. The argument which followed almost cleared the benches especially since Dee and their defensive back were still exchanging words

After the excitement calmed down, we not only lost Dee but the argument about the catch. On our next series, with Dee out of the game, the defensive back decided to concentrate on me. The first couple of plays, he flattened me before I could leave the line. The next time we got the ball, one of our larger linemen said something to me in the huddle. He said “Okay it’s time we took care of that bastard guarding you. Line up way outside. Delay two counts and come back toward the center. When he goes to block you ….hit him high. I’m going to hit behind him low”. It worked a little too good. When he saw me coming at him like I was going to block, he squared around to set his feet. Just about that time our lineman, who was about 6’2” and 275 pounds threw what amounted to a clip. They carried him off the field on a stretcher. His leg was not broken but his wrist, when he tried to break his fall was fractured. We got a fifteen yard penalty.

The time on the clock ran out with the teams tied zero to zero. Since this was the final game with so much riding on it, we had to play off the tie. The Flag Football Rules Book came out and after a conference with the referees, both coaches (and I might add Lieutenant Colonel Hobart was in that huddle), a decision was reached. We would flip a coin. Each team would have the ball for one series. If only one scored that team would win. If both scored each team would get one more series. If neither scored, the team making the most yards in four downs would be declared the winner.

We won the toss and elected to kick off. Actually it just meant they got the ball first as the rules said each team will start from their own ten yard line and have four downs to score or advance the ball. First down markers were ignored. You only would have four downs and yardage counted in total only. An interception, unless returned for a touchdown only meant the other team lost the ball and would still have the yards accumulated before the turn over. There were no fumbles. If the ball hit the ground it was dead. They gave each team about ten minutes to talk things over and plan strategy.

The rules were so funny that a certain amount of planning would be required. There was an advantage if you had the ball first and scored. Then the other team had a lot of pressure on them. On the other hand you needed to have some yardage just in case you did not score. Our quarterback had made the decision on the toss of the coin. He told us his thinking was if he won the toss he wanted them to have the ball first. He said "They are so arrogant, I'm betting they go long at least twice trying to score. We have kept them out of the end zone all day, we can still stop them".

On First down, they threw for nine yards. On Second they threw a long pass for the end zone. Lucky for us their quarterback seeing his man wide open got excited and overthrew him by ten yards. Our defensive backs forgot that making a First down did not count in the playoff and they were playing up close because it was Second and one. They were expecting a short pass. Our coach tried to call time out but he was too late. On Third down their quarterback ran around right end and picked up twenty-five yards. He almost broke it all the way. A diving grab at one of the flags saved the touchdown. This brought them to Fourth down. They had gained a total of thirty-four yards. They called time out to discuss their final play. They had blocked so well on the running play and it had gained so many yards, I sorta expected them to try it again to the other side. Coach Sullivan had our defense in a huddle also. His theory was that they had been a high scoring team all year. He did not think their ego would let them admit they couldn't score on us at least once. Fortunately he guessed right but as the play started out it looked like their quarterback was going to the other end as I had thought. He stopped short of the line and lofted a hail Mary in the direction of two of his receivers. Our defensive back went up with them and managed to bat the ball to the ground. Hobart was screaming interference but it was not called.

Now it was our turn. All we needed was to get more than thirty-four yards and we had four downs to get them. Our Quarterback, Airman Doss came into the huddle and said "Alright guys we don't have Dee but we are going to run a play which has worked well all season long ….. "TWO AND THROUGH". All our plays had names where it would be easy to remember them. Doss had them all diagrammed on index cards and he would pull them out each time we went into our huddle. He did not have them out now. "TWO AND THROUGH" was my play. I did not have another where I was the only receiver. There were several where Dee and I ran crossing routes or where he or I would decoy but that was my only exclusive play. I was to line up at left end, delay two counts after the ball was centered and then break directly across in front of the center. The ball was always a bullet pass and Doss would throw it hard. We had used it all year on short yardage, usually where we needed four or five yards for a First down. The ball was snapped. I counted slowly ..one..two and broke across center. The ball was there fast. I held on and gained nineteen yards. We went back to the huddle and Doss patted David Witherspoon, our tallest player on the head. He was 6’6” and normally transferred to Headquarters Squadron during basketball season. Doss said “David everyone on their team knows you and they know you can out jump them. They’re going to see me patting you on the head and think you are getting the ball. Go long yelling for me to throw it high and take them with you. We are going with the same play “TWO AND THROUGH”. I probably delayed longer than two counts because when David ran out screaming throw it high, three of their defenders all converged hollering “Cover David, Cover David”. I ran the pattern and Doss threw another hard pass right in my gut. Someone dove and grabbed a flag, knocking me off my feet but not before I had gained sixteen yards. We went back to the huddle. Headquarters had not even realized we now had them by one yard. They were already lined up on defense awaiting our next play. Doss said “Hell we’ve beat those sons of bitches, all we’ve got to do is punt”.

In Flag Football there is no rushing the punter. All you have to do is say Free Kick and punt. Doss raised up and said "Free Kick we're going to punt on Third down". This brought a scream of protest from the other side. Their coach ran on the field saying "They can't punt on Third down in a playoff. They have to have Four downs like the rule says. That way they stand a chance to lose yards". Our coach who by now was in the middle of the fray said "You dumb-ass that's exactly why we are punting on Third down. We give up the ball and it ends at the line of scrimmage. We win by a yard". The referees were losing control fast. By now Hobart was out on the field yelling that we couldn't do that. "The game isn't over until they run all four plays like we did". I tried to tell Doss that it didn't make a dam. All he had to do was throw two long incomplete passes and we would still have thirty-five yards. He and I were not in the middle of the argument. In fact our whole team stayed back and let Coach Sullivan fight it out.

The referees got together on the sideline. Hobart was waving the rule book and they did study it for awhile. Much to the credit of the Head Referee he asked Hobart politely to stay out of it and let him make his own decision. The rules for a playoff were not clear but all the referees agreed that a team could choose to punt on any down. The head Ref said "The team with the most yardage is the Twenty-Fifth Instructor Squadron, so I declare

them the winners”. I wish Doss and Witherspoon had not decided to hoist me on their shoulders and carry me off the field. This only seemed to upset Lieutenant Colonel Hobart all the more.
That night the entire team ended up partying at the NCO Club until it closed. I think I paid for about five pitchers of beer but it was worth every penny. Witherspoon had helped carry me off the field. I returned the favor by helping carry him back to the barracks. He tried drinking a beer for every point he had scored during the season; he almost made it.

MONDAY MORNING STAFF MEETING

Sunday was a time to reflect, to think about what winning that important game would mean. In my case it could mean the end of a short Air Force career. I dreaded the upcoming Commander’s Staff meeting on Monday. Mike stopped by my room. He knew about the letter I had written to Colonel Favor and he had noticed that Favor was at the game and he had been pulling for Headquarters to win. “What do you think he will do”, asked Mike. “I don’t have the slightest idea”, I replied. “If you hear I have been reassigned to Tulle Greenland, you’ll know he wasn’t too happy”. The meeting was set for 9 a.m. on Monday. As you might guess I did not sleep Sunday night. I had time Monday morning to run by my office first. Sitting in the middle of my desk was a large sheet of poster board with stick figure cartoon drawings of what appeared to be an Air Force Lieutenant Colonel pulling the gold bars off the shoulders of a Second Lieutenant. The caption under it said "TWO AND THROUGH ... MY ASS … I'LL TAKE THESE TWO AND YOU ARE THROUGH". It was signed by our coach and several teammates who lived in our barracks. Sergeant Summers had added "GOOD LUCK … IT'S BEEN NICE KNOWING YOU". I tried to get a laugh out of it but I will admit panic was starting to set in.

Walking into the meeting, I felt that all eyes were on me even though I was pretty sure most of those in attendance were not aware of my letter or even the outcome of Saturday's game. Most of the people who came to Colonel Favor's staff meeting could have cared less about the Commander's Sports Trophy and besides most of them were in Headquarters Squadron. Colonel Favor started the meeting by saying "I want everyone to congratulate Lieutenant Capps. His football team pulled off an upset of Headquarters over the weekend. Also I want to tell you about a letter I received". My heart was in my throat. Don't tell me he is going to read my letter. He definitely had my letter in his hand holding it up. He said, "I must say this is the most unique letter I have received in my Air Force career". He looked directly at Hobart as he spoke. "I'm not going to read the letter but I will tell you what it means to me. When I was in college, I ran track. I was not good enough or should I say fast enough to be one of our stars. I did, however, play an important role on our track team. When running the mile, I was one of the pacesetters. My job, as that of other pacesetters was to start out a race running fast and try and get the opposing team to try and keep up. Hopefully they would wear themselves out as I did, so our regular track stars could then easily win the race. This letter is about a pacesetter". He paused and then spoke loudly "Nothing gives me more pleasure than to occasionally see a pacesetter win". Not another thing about the letter was mentioned. The only people in the room who had any idea what Colonel Favor had just said was Hobart and me … and I was the only one who enjoyed it.

Colonel Favor was a wise person. Actions like this would eventually propel him to become a General. Several years later I would have the opportunity to visit General Favor in the Pentagon. He had not changed. He was still nice, easy going and a role model for junior officers. He welcomed me into his office like an old friend. But even in the Puzzle Palace of the Pentagon, I noticed an outer office which had a small sign on the door with the title "Aid To General Favor … Colonel Frank T. Hobart". Note: That was not Lieutenant Colonel but Full Colonel. For those of you not familiar with the military, when you speak to a Lieutenant

Colonel, as a courtesy they are addressed simply as Colonel. I can tell you that most of the Junior Officers on base all wanted to say Lieutenant Colonel Hobart when they spoke to him but none would ever dare.

MANY HOURS IN A DAY

Since I covered sports from all my time stationed in Amarillo it probably sounded like that took up all my after duty time. That's far from the truth. Other than softball and football season (and they only consumed a couple of afternoons or nights a week), there was lots of free time. As I said I wanted to stay busy so I quickly gained the reputation of accepting more and more extra duties. Some I volunteered for and others were assigned because they knew I would get them done. One thing about the Air Force, if you are willing to work they will pile on everything you can handle.

Another reason I got assigned extra duties outside the squadron was because I became well known among senior officers around the base. My dad was active in our Masonic Lodge back home. He told me about an organization called National Sojourners which is an American patriotic organization of Freemasons who have served or are serving in the U.S. Armed Forces. He said "Why don't you see if they have a chapter on your base". I did and a Major in our squadron was a member. He was surprised that at my age I was already a Mason. He invited me to join at their next meeting. When I walked in he started introducing me around. There were about 15 or 20 members (most in civilian clothes although several who had come straight from work were in uniform). I am not good at remembering names but at each introduction, I caught the ranks. It was Major this, Lieutenant Colonel that or Colonel (a name I recognized as the Base Commander). It was obvious I was not only the youngest (by far) but the only Second Lieutenant. They were all asking about how I had become a Mason at such an early age. I told them about my dad and grandfather. I was accepted immediately. Another Colonel (and not the Base Commander) seemed to be running the meeting. The night I attended the meeting they were electing officers for the year.

The first to be voted on was Secretary. The Base Commander said, " I think Lieutenant Capps was sent to us for a reason. I nominate him to be Secretary". The Major who invited me said, " I second the nomination and make the motion he be elected by acclamation". Someone seconded that and I was elected Secretary to the applause of everyone in the room. The next officer was Treasurer. Again the Base Commander said,
" Well Lieutenant Capps has done such a good job as Secretary, I think he should be our Treasurer

also, so I nominate him". By now everyone was laughing, someone seconded the nomination and I became Secretary and Treasurer before I knew what was happening. Before we left that night I was handed all the books showing the minutes from previous meetings and the past Treasurer's reports. We were to meet again the next month.

By the time of the second meeting I had studied all the books and found out all the names and ranks of the members. It was like a who's who of ranking officers on the base. The membership was made up of mostly Colonels, Lieutenant Colonels, Majors and two Captains. There were no other Lieutenants (either Second or First). When I was called on to read the minutes of my first meeting I read off the names of who had been elected. I started with the President and worked my way down. When I got to Secretary and Treasurer I said, "Lieutenant Capps was elected with a railroad job of the Base Commander pulling rank over everyone in attendance". This broke everyone into laughter. Then I added "Okay I am also going to give the Treasurer's report. Several of you are behind in your dues (most of them were) and if you think I am going to be intimidated by your rank, think again I will bug you until you pay". Before we adjourned I had collected from everyone there including the Base Commander (who said before he left, Major Angus told me about you). My peer group of Lieutenants were always surprised when we were at the Officer's Club and Senior Officers from all over the base would speak to me and call me by my first name.

My reputation had come to the attention of Colonel Favor well before football season. I had mentioned in that commentary about me representing the squadron as Acting Commander because Major Angus had been promoted to Lieutenant Colonel. I like to think I actually had something to do with that. Colonel Favor had called me to his office one day. He wanted to congratulate me on the job I had been doing and wanted to know why I was not going to make the Air Force my career. He often did these pep talks with junior officers, who on their ID Cards were shown as Reserve Officers with a date of separation instead of Regular with the date of separation shown as Indefinite. He said he had been getting reports of my outstanding work and he wanted to know why I wouldn't consider a career in the Air Force. I told him that I signed up for four years and that I planned on doing that to the best of my ability but that I had a degree in Advertising Design and I wanted to get out and do something in that field. I then went on to say the main reason was that most all high ranking officers I had come in contact with were rated, either pilots or navigators and even some of those were overlooked. I said, "The sharpest officer I have met since I have been in service is Major Angus. He was decorated as a pilot by two countries, having flown with the RAF before we got into WWII. He has more time in grade than most Majors and he has not been promoted to Lieutenant Colonel and I can't understand why". I could tell that last remark took him by surprise. He said "You know, our Base Commander told me that you are not intimidated by Senior Officers and say what you think. I see what he meant. I can promise you I am going to look into the situation with Major Angus".

About two weeks later everyone was congratulating the Major on his promotion to Lieutenant Colonel. He whispered "I think you may have had a hand in this". He and I had become close. We worked well together. The atmosphere of Bill and Ralph carried over to our whole staff in the Squadron Headquarters Orderly Room. Everyone got along and did their jobs. Me taking over as acting Commander was smooth because nothing really changed. I had a few more papers to sign but Sergeant Summers and the rest of the staff kept everything running. I said the Major and I became close. He and his wife Ruth invited me to their house for dinner a few times. On one occasion Ruth called me aside and said "You and Ralph have a special relationship. You are like the son he never had". They had a teenage daughter who would run and hide when I came over because they failed to tell her a single Lieutenant was coming for dinner. Anyway Ruth went on to say “Because you two are close, sooner or later you’re both going to have too much to drink and Ralph is going to bring out his mom”. This was somewhat confusing until she told me his mom had past away several years before, had been cremated and her last wish was for her son to spread her ashes from a plane while flying over California. This broke me up but sure enough, one night it happened.

We had started at Happy Hour at the “O” Club, then continued drinking at his house. He got to talking about responsibility and out came the Urn with his mom’s ashes. He said that he was supposed to spread them over California but they got transferred to Amarillo before he got the chance. He knew it was a responsibility he had not taken care of but every now and then he would drink a toast to his mom and promise to eventually take care of her last wish. His next door neighbor (another Major) was with us sitting on the front porch. We all drank a toast to his mom and probably toasted a few other people before we called it a night. I bugged him about taking care of his mom’s ashes until I was transferred from Amarillo. Eventually Ruth wrote me to say they had gone back to California, rented a small plane and took care of her final wish.

THE SQUADRON COMMANDER'S POSITION

The 25th Instructor Squadron Commander's slot was for a Major. A Captain can fill a Major's slot but if it is filled by a Second or First Lieutenant it can only be as Acting Commander for up to ninety days. After that period usually a Major is appointed. Well when Major Angus became Lieutenant Colonel I assumed command until a new Major could come in. It was almost the end of three months when Major Stephens took over as Commander. He was also a pilot with twenty years active duty. He had never been a Commander and he had always wanted to be one. He came in "gung ho" and wanted to do every thing by the book. He wanted to be much more formal than the style we had become accustomed to. He was a nice enough guy but way too serious. That first week we were visited by the Inspector General's Office. This was usually a once a year inspection, normally by a Colonel. He would check all the records and everyday actions in each squadron. Most every Squadron Commander dreaded the inspections. Since Major Stephens was brand new, he had to rely on me to deal with the inspector, especially since I had been running the squadron as Commander. When the Colonel came in our Orderly room popped to attention. Even Major Stephens seemed a bit nervous when he met the Colonel. The Major explained that since this was his first week that I would have to be the main contact for our inspection. The Colonel came in my office with a clip board in hand and sat down in front of my desk. He said “Sit down Lieutenant and let’s get acquainted. Where are you from”? When I said Birmingham, Alabama he said “No shit, you know where I’m from”? I replied “Well since you are going to be inspecting us, I hope you are from Alabama”. He said “Yep, I’m from Mobile, I bet we are going to hit it off”. We chatted for awhile about our state. He had gone out of state to college while in service but knew quite a bit about Auburn. In fact one of the ROTC instructors was a relative of his. I did not remember him and told him my commission had come from OTS and not ROTC.

Finally he said we should get on with the inspection. He referred to his clip board and said let’s first go over your records. I blew him away when I said “Okay you want to see the records I keep for inspectors or the ones I actually use to get the job done”. He said “Boy that’s a new one, show me what you mean”. I proceeded to point out how some of the stuff we were required to keep was useless to our everyday jobs. Much of it was what Sergeant Summers had first told me about. Some people create reports just to justify their jobs. I kept the files because they were required but we never used them. The Colonel took copious notes. When we got around to inspecting the various charts we had to keep on our walls, again I threw him for a loop. I showed him my chart which explained how to read all the other charts. By looking at the one in my office you really did not need to look at any of the others. I had brought all of my art stuff from school so I had pens to hand letter signs and charts and they looked professionally done.

To make a long story short the 25th was the only squadron to get a perfect score in every aspect of the report from the Inspector General. Of course at Colonel Favor's weekly staff meeting Major Stephens received all the credit. Lieutenant Colonel Angus knew the truth because Sergeant Summers had told him and I am sure he passed that on to Colonel Favor. Later in the week I was pulling extra duty as Officer Of The Day for the whole base. A different officer pulls this duty each day staying in the Base Command Post for 24 hours. You don't get much sleep but it depends on what happens on base while you are on duty. Someone stays by the phone (normally an NCO) at all times. There was a bunk where the officer could often grab a nap. About 1:00 am the NCO woke me up to say there was an officer, a Colonel who was drunk in the lobby. When I walked out, there was the Colonel from the Inspector General's office. "He did not recognize me at first and said, slurring his words "Lieutenant, call me a taxi". I said "Okay Colonel you are a taxi". He looked up and said “Oh hell it’s that smart ass Capps from Alabama”. I said “Yes but it is that smart ass Capps who is going to get you back to the Visiting Officer Quarters before you fall and get hurt”. I called the motor pool and got an airman to drive him back to the VOQ and make sure he got back to his room okay. The next day he came by my office to apologize. He said that he was glad it was me on duty as Officer Of The Day. He stayed and chatted for about thirty minutes. After he left Major Stephens came in and asked why the Inspector had returned to our squadron. I never told him what had happened. I just said that he wanted to confirm some things we did for his report. He actually did make recommendations for other squadrons about simplifying some of our reports based on what I had shown him.

Commander Again

Weird things did happen to me in the Air Force. Right place right time or wrong place wrong time they just happened. Major Stephens was almost too conscientious about his new job. He wanted everything run by him first. All of which we had been doing routinely. Sergeant Summers was not used to being checked on nor were any of our staff. They were coming to me saying they could not get anything done because the Major wanted to see everything before it was done. He had only been in the job about three weeks when a real emergency came up. With a squadron as big as ours, with all permanent party personnel, big problems were a common occurrence. It was a weekend on Saturday night. Normal procedure if a squadron has a problem of any kind a chain of command is followed to try and solve the issue. The first call is to the Sergeant Major (Master Sergeant Summers). If he is not available or cannot handle the situation the Administrative Officer (me) is called. Same thing … the next called would be the Squadron Commander (Major Stephens). Well that particular Saturday, the Sergeant and his wife had gone out of town, I had a date and did not return to base until after 2:00 a.m. so the Major was called. Two NCO's from our squadron, one white and one black had gotten in a fight at the NCO Club. The Air Police were summoned to break it up.

When I arrived at the BOQ there was a note on my door to call the Base Hospital. It was from The Officer Of The Day and it said "There is an emergency, ask for the Hospital Commander" He happened to be a Colonel who I knew from The Sojourners. When I called he said "You need to get down here and wear your uniform. I will explain when you get here". Needless to say I dressed and got there in a hurry. He met me at the door. The two NCO's both of whom I recognized were there along with the Provost Marshal (who was a Lieutenant Colonel who I also knew). The Hospital Commander said "Major Stephens responded to a fight with members of your squadron at the NCO Club. He apparently got very excited and started yelling at the NCO's involved and had a heart attack .... and he died". I guess I stood there in shock. He then added "You need to cut orders assuming command of the squadron and tell the Provost Marshal what you want done with your NCO's. When I talked to him he said that one of the NCO's had given the Major mouth to mouth and tried to save him but he was pronounced dead when the ambulance arrived. He said he thought they had both been through a lot and did not think they should be punished further. He said the argument started after too many beers and it was actually resolved before the Major arrived. They were both really upset because they thought they were responsible. I just dismissed them and told them to report to my office Monday morning. I went back to the office typed up the order myself again assuming command of the 25th.

Early Monday the Commandant Of Troops, Lieutenant Colonel John Parker was waiting for me in my office. He was who Major Stephens reported to (whereas Major Angus and I as Acting Commander used to report to Colonel Favor). Colonel Parker said "As Acting Commander again you now report to me. Your first assignment from me will not be pleasant. You need to read the regulations on how to handle the body of Major Stephens. I can't help you because I have never been faced with something like this”. I told him that before I left the hospital I had called the Base Chaplin (woke him up) and he had said he would be responsible for notifying Major Stephens wife. She had stayed in Arizona in Phoenix. Major Stephens had been stationed at Luke Air Force Base and was planning on retiring in a couple of years. He wanted the experience of being a Squadron Commander and got that chance at Amarillo. His wife and their only son stayed behind because that was where he wanted to retire. Colonel Parker said “Sounds like you are already handling everything. If I can help you just let me know”.

The first person I called was Lieutenant Colonel Angus. I figured with all his experience during WWII he had gone through something like this. He said that the Air Force had explicit regulations which explained everything which had to be done. He said a Summary Court Officer would be appointed for the purpose of disposing of the personal effects, property etc. for the Major. Funds would be provided for the packing, crating and transportation to his next of kin. That afternoon Colonel Angus called to tell me a Major had been appointed but he had asked that the Colonel and I assist him since we both knew Major Stephens. We actually ended up doing the job.

Before we met with the Summary Court Officer, Colonel Parker called to say that he had arranged for the body to be flown to Arizona and that I and Colonel Angus would accompany the body for the funeral. He said there would be room on the plane (a C-130) for around 12 passengers and that I should decide who of the Major's friends or co-workers should go. Sergeant Summers and I came up with the list. Major Stephens lived in the BOQ as I did but he was in a different building. I had only been to the Major's room on one occasion but I remembered that he did not have a lot of extra stuff. The Air Force is very specific on what can be sent. No alcohol, no pornographic material, no firearms and nothing which might embarrass the next of kin. The Major did have a bottle of Jack Daniels, a few copies of Playboy and a 45 automatic pistol. Colonel Angus said that since we would be going to the funeral we were going to ignore the rule about firearms because he thought the Major's son when he got old enough would like to have it. Also technically we would not be ignoring the regulation because we were not shipping it…. but hand carrying it to his widow.

One of the passengers Sergeant Summers suggested was an NCO who had known the Major at a previous base. He was a Senior Master Sergeant and his job at Amarillo was seeing that all students going through Tech School were inspected and looked and did their best in every aspect of their stay at Amarillo. He taught drill, he taught spit and polish, he made their lives miserable. As you might guess he always looked like the guy on the recruiting poster. He was a perfectionist. When we arrived at Luke Air Force Base he demanded that he meet with the airmen assigned to perform the military funeral which had been arranged. He made them practice marching with the casket, and folding the flag. They did it several times until it met with his approval. He even had the bugler practice taps before we left for the cemetery. The service was beautiful. His wife and family members really appreciated all of us coming. With many of his friends from Luke also in attendance it was a large crowd. We flew back that same night. I was gaining experience far above what I imagined when I joined the Air Force.

FIRST HAND EXPERIENCE … THE BEST TEACHER

Again I find myself Commander of one of the largest squadrons on base and all permanent party. I think I said before that we had 44 officers in the squadron. All of them outranked me except my friend Mike and Dee from the football team. Normally the Commander of a squadron is referred to as "The Old Man". I had two Lieutenant Colonels in the 25th. One of them who was always at the Officer's Club with his wife and friends would call me over to their table and say "Let me introduce you to my "Old Man". He is my Squadron Commander". It got a good laugh even though he did it several times to the same group. He did pay me a compliment by saying "If you have a problem, call Lieutenant Capps, if he can't handle it himself he will find someone who will". Another time he was with a large group of Senior Officers. He called me over and introduced me (again as his "Old Man"). He told them about a recent practice alert we had where everyone had to go to their assigned shelter. It happened to be in his building. I had gone over and took a clip board so I could write down a note if someone from my Orderly room failed to go to the shelter. The Colonel came up and said "Lieutenant I got orders that I have extra duty as the Shelter Warden and I don't have the slightest idea what I am supposed to do. You are on the same set of orders as the Alternate Warden. Do you have any suggestions"? I had handed him my clip board and said "Just walk around with the clip board and everyone will know you are in charge and think you know what you are doing". When he told the story, all the officers cracked up. Then he added "Lieutenant Capps does not always go by the regulations but he knows how to get things done. My reputation among Senior Officers was growing.

Most of the time that's what I did …. find someone who could solve whatever problem came up…..and boy there were many of them. The day after we returned from the funeral Sergeant Summers came in and said “While you and I were gone we had an airman in our barracks attempt suicide. He could not wait to give me the details. He said that our Airman First Class who was in charge of our Orderly room was called to the bathroom where they found one of our airmen on the floor with a bottle of Cadet Ink lying next to him. Cadet Ink was used to polish shoes and boots and was marked POISON on the label. The airman had black ink on his mouth and was rushed to the hospital. Summers and I headed that way. Again I was met at the door by the Hospital Commander. He said “I might as well get you an office here”. He went on to say that they had pumped the airman’s stomach the night before and there was really no sign that he ingested much if any of the Ink. Summers immediately said “That little son of a bitch is just trying to get out of the Air Force” he said “I know him and he is just a flake, he’s always complaining and trying to get out of work”. Remember Sergeant Summers kept the Duty Roster so I was pretty sure this kid was on his list. The Commander said “He is being released this afternoon, I’ll tell him to report to your office”. On the way back to the office Summers said “In the old days I would take somebody like that out back of the barracks and beat the shit out of him”. He was from the “Old School”.

When he showed up I could hear the sergeant screaming at him outside my office. He finally got the airman to admit (after he started crying) that he faked the suicide. He only admitted it after Sergeant Summers walked to the window, opened it and said "You want to kill yourself jump out this damn window, that concrete will do a better job than Cadet Ink". That's when he started to cry. I hate to admit it but the kid was from Birmingham. We discharged him under an unsuitability clause. When I got out of service I ran into him selling suits in a men's wear store. He tried to act like he never met me.

Someone was always in trouble. We just had too many people and the law of average said it would happen almost once a week. Colonel Favor started a campaign to stop Tech School from having so many off base traffic tickets. Speeding, running red lights and stop signs, DWI, fender benders, loud mufflers, you name it and someone did it. My old buddy Lieutenant Colonel Hobart came up with an idea which Colonel Favor approved. Hobart started his own Traffic Court. Every Monday morning at 6:00 a.m. in the Staff Meeting Conference room the offending person along with their Squadron Commander had to appear before "Judge" Hobart. Different Student Squadron Commanders would often have to appear but I became a regular. I never missed a meeting. Hobart's theory was to put pressure on the offending person by suspending their on base driving privileges. Making their Commanding Officer have to also appear with them at six o'clock in the morning would also embarrass them and make them think twice about committing another traffic offense off base.

One Monday, Hobart went too far. He was beginning to feel his power as a Judge and handing out punishment. I had an Airman First Class who lived in the squadron headquarters barracks. He had a 1957 Chevy convertible (bright red). It had extremely loud mufflers. He had been stopped multiple times from everything from speeding to running red lights to you name it. It was his third week in a row to accompany me to "Traffic Court". Each week I probably had at least three squadron members who had received an off base ticket. As soon as Hobart saw my airman for the third week in a row he blew up. He said "Obviously suspending your on base driving privileges is not enough, I am ordering you to turn over your drivers license". The airman said "Sir I don't have it with me". Hobart yelled "Well I am hereby ordering Lieutenant Capps to take your license and hold it until further notice".

When I got back to my office later that day the airman's drivers license was on my desk with a note from him saying that he thought it was illegal for the Air Force to take his license and he was calling his dad about it. I had heard that his family was wealthy but had no idea who they were. I decided that I needed to call my buddies in the JAG (Judge Advocate General) office. I had met one of the lawyers my first day at Amarillo but by now I knew everyone in that office and they knew me. I asked if I were given an order which I thought was wrong or illegal did I still have to obey it? They broke up when I told them what it was. They were all familiar with Lieutenant Colonel Hobart. They told me that basically the answer was "Yes" unless it endangered either my life or that of someone else. They added "Just to cover your ass, put the order in writing and show who gave it to you". I did, and later that week the squadron received what was called a "Congressional". If anything was received from a member of the House or the Senate, it had to be answered within 24 hours. The JAG office also received a copy. It hit the fan. The airman's dad was a big donor to a United States Senator. Needless to say he got his license back, I got an "atta-boy" from the JAG office for giving them a heads up and Hobart hopefully at least got chewed out.

As I said I got exposed to many things fast, especially as Acting Squadron Commander. I received a call from the JAG office. I now had conversations with someone in that office almost daily. A young Airman who worked in our Aircraft Maintenance School had not reported for work for a week. When his supervisor checked with his roommate, he said he had not seen him since the Friday before, so he had been gone nine days. He was reported AWOL (absent without leave). He lived in my other building so I had someone go over and leave a note for his roommate to come to my office the next morning. After I questioned him I was convinced that the Airman had gone to Arkansas to see his girlfriend. The roommate said they had been having trouble and he thought she had a new boyfriend.

I reported my small investigation to the JAG office. They said, "Well they probably won't really look for him until he has been AWOL for thirty days because then it would be desertion". In time of war this would be a serious crime. Even if not, it still could involve a court marshal or even jail time. The lawyer I talked to said if and when the airman was caught, if he had his military ID on him, he would probably not be charged with desertion. I let his roommate know because I thought they had been in touch with each other. When he was caught and returned he was only charged with being AWOL. He received an Article 15, busted from Airman First to Airman Second and received no time in the stockade (all because he still had his ID). The Stockade, however I would become familiar with. The first time I assumed command of the squadron when Major Angus became a Lieutenant Colonel I got a call from the JAG office informing me that I had a member of my squadron serving time in the stockade. There was a regulation which said Squadron Commanders, if they had anyone from their squadron in lock-up must personally visit them once each month to check on their well being. Oh great I did not know we had anybody there. Checking with Sergeant Summers he said "Yeah, he's been in and out of the city jail downtown and he finally screwed up enough on base to get stuck in the stockade here. He used to be a Tech Sergeant and was a fair instructor until the bottle got the best of him. The Major had him sent to the military hospital in San Antonio to dry out but it didn't work". I got his name, saw all the offenses he had committed over the last year, and went to pay him a visit.

The Senior Master Sergeant in charge of the stockade was pretty sharp. He told me those visits, even though required, didn't amount to much, mainly because the prisoners did not want to see any officers ….. especially a Second Lieutenant. Even so, I told the Sergeant to let me see and talk to him. My visit lasted about three minutes. I asked him how he was doing and if he needed anything. His reply was " If you can get me outta here do that … if not piss off and leave me alone". I did. The next month, I came down , signed the register showing I did my duty and left. Same for the following months.

The city jail was mentioned above. Over my eighteen months stationed in Amarillo I became well acquainted with the City Jail downtown. In fact the Amarillo Police Department Lieutenant and I became friends. One of the extra duties of all junior officers on base was to ride "Town Patrol" with the Air Police. Your name would only come up about every six months. This was so we would know what the city police had to contend with each weekend when student airmen and even permanent party personnel caused problems. A lieutenant would ride with an Air Policeman in one of their vehicles around downtown Amarillo on Saturday night. We would leave the base around 6:30 in the evening and return around 1:00 or 2:00 a.m.

We would hit most of the night spots, many which were off limits to airmen. We would always start out and end up at the city jail. This duty was mostly assigned to Second Lieutenants but occasionally a Captain or Major would volunteer. Most of my peers hated pulling "Town Patrol" but I kinda enjoyed it and often swapped one of my extra duties and took theirs. One of my first nights I was riding with an Air Policeman who also taught Judo on base. He looked like he lifted weights on a regular basis. He knew most of the Amarillo cops by name. We had just left a bar when we came upon flashing lights. An arrest was being made of a guy who had thrown a brick through a jewelry store window. He had started to run with a sack full of stuff he had stolen from the window. Apprehended, he was sitting in the back of the patrol car. The policeman recognized the Air Policeman and said "Why don't y'all follow me back to the station and the Lieutenant can see how we book a prisoner and get a tour of the jail. It was located upstairs above the station offices.

We went in the office and I met the Lieutenant for the first time. He said "I'll come upstairs in a few minutes and give you a tour of our jail". The prisoner, the arresting officer, the Air Policeman and I all got on the elevator. It was not very large. The prisoner had been extremely polite, apologizing, and said he had too much to drink and was going peacefully. For some reason the officer had taken the cuffs off the guy. He was still saying how sorry he was until the elevator door opened on the top floor and he saw the bars. He started swinging. He drew back to hit the policeman. When he did his elbow caught me on the side of the head and I went down. Within a blink of the eye, the Air Policeman had him on the floor halfway out of the elevator … in a choke hold. I got my tour from the Lieutenant while holding an ice pack on my head. That's probably why we became friends. He always remembered my first visit. I would go back to the City Jail many other times and not on "Town Patrol". It seems the 25th

Instructor Squadron had another individual with a serious drinking problem and he happened to work in my Orderly Room. He was a tiny frail looking guy and older, especially for an Airman Third Class. My early experience with him when I was the Administrative Officer had been good. He was in charge of all the furniture, file cabinets, typewriters and everything issued to us in the Orderly Room and both of our buildings. He kept the computer printouts and was an expert on inventory. Keep in mind this was 1963. Computers were huge, kept in a central location and all we saw was the paper trail they produced. He was a problem I inherited from Major Angus.

Sergeant Summers explained that at one time our airman had been a Tech Sergeant. For over 18 years he had an outstanding record. When his drinking started getting worse he kept getting busted in rank ….. now all the way to Airman Third Class. For this book I will call him Airman Charlie Briggs. Summers told me that Major Angus was determined he was going to keep him out of trouble until he had his twenty years so he could retire. He felt the Air Force was partly to blame for the alcoholism. Summers said "I guess we should have told you about him sooner" His main problem with getting in trouble was he lived off base with his wife who was twice as big as he was. They both drank and would eventually end up in a fight. Being much smaller than she was…...he always lost. The most important part of the problem was, he lived directly across the street from the Amarillo Police Station. My first call about him came from the Lieutenant at the Station. Airman Briggs and his wife had gotten into one hell of a fight. Neighbors called the police. Briggs was arrested and taken across the street to jail. He was bleeding but had been patched up by the nurse at the jail. When the Lieutenant asked how he got cut, he said he fell off a chair changing a light bulb. He had been arrested by them for fighting twice before so they knew him. This time when he was asked for the name of his commanding officer, my name came up. I went down to the station. He was released into my custody. I took him back to the barracks, had Summers find an empty room in our barracks and confined him there until I had a better plan.

I called Lieutenant Colonel Angus at home and said "Thanks a lot … you never told me about Airman Briggs". He apologized and said he hoped I would take up the cause and help him to make it to retirement. He only lacked about 5 more months.". I asked for suggestions and together we came up with a plan to send him to the hospital in San Antonio to dry out". I told the Colonel that apparently they were not too successful with the last guy he sent there. I told him about my experience at the stockade the last few months. He said "Well he was completely different, just a troublemaker from the start. I also forgot to tell you about the monthly visits to the stockade. All I used to do was stop by there once a month, The stockade is only 2 minutes from the Base Commissary so I would just stop by on my way there and sign the log. The prisoners don't want to see you anyway as I am sure you found out".

Meanwhile back to Briggs. Until we could get him sent to San Antonio we decided he needed to stay on base in the room in our barracks. One of our airmen took him home to pack up his clothes and return to move in. Everything went well for a week. Being in our building, he was first at his desk every morning and even had Sergeant Summers coffee ready. He kept saying that he didn't think we needed to send him to San Antonio. We did not agree and he was to be sent as soon as possible. Two days before he was to depart, I got a call around 10:00 at night. Briggs had stolen a bottle of hair tonic from someone in the building. He apparently drank it because it had alcohol in it and walked the six blocks to the main gate where he was stopped because he was in his underwear. When the Air Police asked where he thought he was going he had said he wasn't sure.
I really don't remember how long they kept him at the hospital in San Antonio. I only remember that when he returned he looked great and only had about two weeks until he could retire. Lieutenant Colonel Parker (Commandant Of Troops), who

was now my boss had been checking regularly to see how Briggs was doing. I told him he was back, living again downtown and had been coming to work on time and doing a good job. That same night I got a message to call the Lieutenant at the Amarillo Police Station. Sergeant Summers rode with me downtown. It seems Briggs had been arrested for battery. He and his wife got into another fight. This time he won by busting a lamp over her head. I told Summers that it did not look like we could get him through even two weeks so he could retire. I asked Summers if he knew if Briggs had any leave left. He said that he was sure he had at least some. I asked the Police Lieutenant "How long can you keep him locked up"? He said "How long you want me to keep him". I said "How about 13 days". That's how I solved the problem and how an Airman Third Class almost retired from the Air Force. I said almost. He did get to retire but when Parker heard my plan (and what Briggs had done) he said, "I will only approve if you give him an Article 15 and bust him to Airman Basic". Now the difference in the monthly retirement check between an Airman Third and an Airman Basic is so small that it would not buy a bottle of booze so Briggs never even questioned the decision.

On Briggs last day Lieutenant Colonel Parker called to make sure everything was going as planned. I could not resist having a little fun. I said

“Colonel we do have one slight problem”. He said “Oh shit, now what”? I said “Briggs has requested a Retirement Parade”. Parker loved my joke and he told it many times. When I wished Briggs Good Luck, I asked him if he knew what he planned to do. He said, “ I’m not sure Lieutenant, I may take a job with IBM”.

Life was never dull. Another incident again brought me in contact with the Air Police, the Provost Marshal and the JAG Office. A Staff Sergeant in my squadron was accused by his step daughter of rape. She was 20 years old and lived with he and his wife in base housing. When I went up to the JAG office, the lawyer who had interviewed both the sergeant and the step daughter told me that he did not believe the accusation. He said that the sergeant had made the rules of the house and the step daughter was rebelling because she did not want to comply. He said, ”I think she has made up her mind to ruin his career. He will not let her borrow his car to go partying and he told her as long as she is living under his roof, she has to obey his rules”. The lawyer said to me “Lieutenant, she’s a doozy. When I interviewed her she came on to me and kept leaning over in a low cut blouse. I did not believe a word she said”.I asked “Okay how do we handle something like this”?

The JAG Officer, who was a Captain said "Well first you have to take him off duty as an instructor. Get his supervisor to give him a temporary assignment until this is resolved. So far the step daughter has not filed any charges off base, only with our office, so at this point let's take our time". That very night I get a call from the Air Police. Keep in mind this was years before cell phones and we did not have phones in our BOQ rooms. There was a pay phone in the hall on every floor. It happened that the one in my building was not far from my room. At night whoever was closest answered it. The Base Psychiatrist (Lieutenant Colonel Derby) who lived across the hall from me grabbed the phone. He got a lot of calls. He knocked on my door and said "Air Police for you".

"Lieutenant could you come down to the main gate? And you better come in uniform. There is a girl here and she is not wearing much. Her dad is in your squadron". I was not sure who they were talking about but had an idea. He did say her dad and not step dad so I was not sure. When I got there she was talking to the Air Police and one of the guards at the gate. They were enjoying the view. She had on a night gown but it was pretty flimsy and you could see through it, especially when she kept walking in front of the spotlight shining on the main gate sign. They told me she had obviously been drinking but they did not want to arrest a dependent and could I just make sure she got home. I went into the guard shack and tried to call her step dad but had no luck. I then called Sergeant Summers but got no answer.

By now another young Airman had walked up and had joined the voyeurs. I said to him “Airman , how about riding with me while I take this young lady home”. I wanted a witness with me in the front seat and I asked her to sit in the back. The two of us drove her to her parents house. The airman kept glancing back. I am not sure they were home but we waited out front until she went inside the house. She never said a word except to give us directions to the house. On the way back to the main gate I copied down the airman’s name and his squadron. I knew his Commander. When he got out he said “Thanks Lieutenant, that’s the best duty I have had since I got in the Air Force”. Thinking she was going to get in trouble for her rendezvous at the gate she decided to drop all charges. The Provost Marshal had made her come to his office after the guys at the gate filed their report of the incident. He told me that after their discussion she told him the truth and he had called the JAG Office. The sergeant went back to teaching and thanked me for my involvement … which really wasn’t much.

RACIAL PROBLEMS AVOIDED

The African American NCO from the football team who I left stranded going to practice that day (when he made fun of my hometown newspaper) lived in the Orderly room barracks. He stopped by my office often and he and I would discuss the racial tensions which had been in the news about Birmingham all the way back to the riots of May '63. I think I mentioned that our single airmen were allowed to choose their own rooms in either of our two barracks. The blacks had segregated themselves all choosing rooms in one wing of our second barracks.

They had done this before I had arrived at the squadron but I told the sergeant that I did not think it looked good. I did not want anyone saying I was responsible since I was from Birmingham. He helped me get some to change rooms and even move some from their building to ours. I asked him to get in touch with me if he even heard anything the least bit racist or prejudicial. One night around 8 p.m. he had me paged at the Officer's Club. I had gone there straight from work and was still in uniform. He said, "You better get to the squadron. There was a fight in the barracks between a white airman and a black and they've taken it outside …. it could get bad. I will try to intervene but I may need you to back me up". I raced down there in my car. The top was down and

I did not park. In fact I drove on the sidewalk right to the middle of our quadrangle, in front of our two buildings.

Might have known. The white kid was my airman who had all the traffic tickets. I didn't think he would want a Congressional on this one. I was told later they had both had out knives but quickly hid them when I drove in. I told both of them and the now crowd of spectators that I did not give a damn how it all started but it was to end now. I told the two to be in my office at 7;30 sharp the next morning. I knew Sergeant Summers would be there by then and he would have positions open on his famous Duty Roster. The NCO who called me said he called Sergeant Summers before he called me but could not reach him. I told him that I was glad he called me because this was exactly the kind of stuff I was counting on him to help me avoid. We never had another problem.

LIEUTENANT COLONEL JOHN PARKER

I have mentioned Lieutenant Colonel John Parker several times. He became my boss after Major Stephens past away. He and I had a relaxed relationship almost like I had with Lieutenant Colonel Angus. Parker was a character and a great story teller. He told about how when he got to the base as a Major. He was a Squadron Commander and he told about when he met his First Sergeant. The sergeant told him that everyday he would have to leave the office around three o'clock because he was the drill sergeant for the Jody Drill Team.

Jody Drill teams were popular on several bases. They were elite marching units that performed in civilian parades and various functions. They marched to a cadence call sometimes called Jody Calls. It is a call and response work song which normally rhymes. The drill sergeant hollers out a line and all the marchers respond. They are all dressed sharply with high boots and white spats. The word Jody came from folklore as it was a guy named Jody who steals your girlfriend while you are deployed.

The NCO told Major Parker " If you become a main supporter of the Jody Team we will soon make you a Lieutenant Colonel instead of a Major" Parker said I told him "Now listen to me sergeant, let's get this straight, I will tell you when you can leave work and I don't need your drill team to help me become a Lieutenant Colonel". He said the sergeant just said "Yes Sir". Parker then said "In about thirty minutes I get a call from the Major General. He said he wanted to welcome me to the base and my new job. He went on to tell me how important the Jody Drill team was to the base and our public relations. He wanted me to be sure my sergeant was allowed all the time he needed to get the drill team in shape. I too said Yes Sir. I called my First Sergeant got me some spats for my boots and went with him to practice. And now I am a Lieutenant Colonel". Several members of my squadron were on the Jody Drill team. After hearing Parker's story I made sure they were never late for practice. For a long time I thought I remembered Lieutenant Colonel Parker's First Sergeant being named Duckworth. In researching this book I found that in 1944 a Private Willie Lee Duckworth, an African American serving in the Army is credited with the "Duckworth Chant" which was the first Sound Off 1..2, Sound Off 3..4 Cadence Count 1..2..3..4. Private Duckworth wrote most of the Jody calls still used today. Maybe he also had a relative in the Air Force.

GETTING OFF BASE WITH HOPS

With the senior officer contacts I made with membership in The Sojourners, many of them were pilots. One night while complaining to one of them about getting off base on weekends he said "Why don't you go to the flight line and see who is getting in flying time". He explained that he and many other pilots stationed at Amarillo who had desk jobs still had to put in hours of flying time. Many times they would do it over the weekend. He said that I could just ask if I could have a" Hop" and go along for the ride He said that he and another pilot were going to fly to Tinker Air Force

Base on Saturday, have lunch at the Officer's Club and fly back and I was welcome to go along. It was the first of many flights I would take to different bases. When I got back to our Officer's Club that night all my buddies did not believe me when they asked where I had been. I said, "Oh I flew to Oklahoma and had lunch". It was amazing how few junior officers knew about or took advantage of that perk.

DATING IN AMARILLO

It was just as tough in Amarillo to get a date as it was when stationed at Lackland in San Antonio. I had told Lieutenant Mike Rex about how I set up the party at the Catholic College and he kept saying we should find some idea for Amarillo. Most of the girls in town were not interested in dating guys from the base. To begin with it was hard to meet them. A popular TV series in 1963 was "Route 66". It was about two characters named Buz and Tod and their adventures while driving in a Corvette in search of adventure. Since Route 66 ran right outside the base Mike and I became Buz and Tod in my Comet Convertible.

We and many other junior officers often ate downtown at a fried chicken place. It was not a KFC but a bit more upscale. They had two females who alternated serving as Hostess and Cashier. One was around 21 and the other about 30. They were both attractive. After several visits Mike managed to get a date with the younger girl. They dated a few times and then one night after we ate there, the girl Mike had dated suggested we all go somewhere after they got off work. We were in Mike's car which had no heater (or at least it didn't work too well). The big deal among young people on Friday and Saturday night was to ride up and down Amarillo's main drag, which was Jefferson Street. I ended up in the back seat with the older girl. At 23 I Was not that much younger but back then 30 sounded pretty old. We were under a blanket because it was cold. She was not opposed to kissing a younger man and I was not paying a lot of attention to riding up and down Jefferson. About the time I no longer felt the effects of the cold weather Mike's date says "What number is your husband's car"? I never lifted my head from under the blanket to say "Number one you never said you were married and number two please tell me your husband drives a cab". She replied "No he's a cop and his car is number 45". Another popular TV series at the time was "Car 54, Where Are You". I guess a dyslexic would flip that to 45. I only flipped out and said "I think we need to head back to the restaurant. Those guys carry guns and they can invent reasons to use them. Besides I know too many people at the Police Station". She continued to say not to worry but we went back anyway. The moral of the story if there is one is "Never Cop A Feel If Her Husband Is a Cop". Big lesson ...No wedding ring does not mean she is not married.

When warm weather was approaching the front page of the Sunday edition of the Amarillo paper had a color photo of a beauty wearing a swimsuit (slow news day). It had attracted the attention of several junior officers. We were at breakfast at the Officer's Club. Sitting with four other Lieutenants, we were passing the photo around. Someone said "Wow we need to figure out how to get a date with her". I said that I bet I could do it. My remark brought out the "Wanna make a bet on it". I said, "Sure just make it worth my effort". One of them said "Okay how about five bucks says you can't get a date with her by Saturday night two weeks from yesterday. And you have to bring her here to the club". The other three guys said that they were in. I took the bets. Now losing twenty dollars would be a lot back then but I took the challenge.

The cover photo had her name and the cut line revealed why she was posing in the swimsuit. What the guys did not know was there was an article on the inside pages which told who her parents were. I looked up their address in the phone book. I wrote her a letter explaining who I was and how difficult it had been to meet local girls. I said that I knew how most of the girls living in town felt about guys from the base, however, she should tell her parents that this letter was coming from an Officer and Gentleman, by act of Congress. Also I was a Squadron Commander for an Instructor Squadron which was permanent party and I was not a student. I suggested that we only meet in the afternoon for a Coke and maybe she could introduce me to her parents. I gave her my phone number.

Bingo …. she called the day she got my letter. Her mom had said she could invite me to their house for the Coke date, and in the afternoon. I did not tell the guys that I had made contact. It ended up being on a Thursday in late afternoon because I said it would have to be after I got off. I arrived at their house around 4:30. I met her and her mom (who was also attractive). Her dad would be coming home around six because he worked at the paper. Now you know how she got on the front page. We sat around and talked and her mom served some snacks and lemonade instead of Cokes. Her daughter and I eventually went out and sat on their front porch. The dad came home a little before six. We talked and I think I met with his approval. They invited me to stay for dinner but I passed, thanking them and said I would love to but I had a meeting that night. Before I left though I set up my date for a week from Saturday. I told her we would go to the Officer's Club for dinner.

Of course word got out that I had won the bet but none of the guys would pay off until I actually proved it by coming to the club. She was dressed as a doll, small but in high heals ...she looked stunning. During dinner those guys and several others came over wanting to be introduced. The four with whom I had the bet had obviously planned the whole thing. They all pulled out five one dollar bills and started dropping them on the table. When they told her why, she handled it well "She said well he told me that you bet ten dollars each so I think you owe more money". Later when we were alone she did not think it was all that funny. She said "You know you could have told me'. We dated several more times but we did not hit it off. I fixed her up with one of the guys who had been in on the bet. When I left Amarillo, they were still dating. She and I remained friends and her dad ended up helping me on a project which I will explain later.

A Major in my squadron approached Mike and I one night at the "O" Club. He was divorced but lived with a girlfriend off base. He said his 21 year old daughter was coming to visit him for a few days and would one of us take her around and entertain her while she was there. He said he was sorry but the only photo he had of her was when she was 13 years old. We both looked at the photo of a freckled face young girl in pigtails. We both said sure we would be happy to but as soon as he left Mike said, "On this one we had better flip a coin to see who gets stuck with her". We did and I lost. We planned a double date. Mike had been dating a cute girl who was the daughter of a Senior Master Sergeant in our squadron who taught in the Missile School. When I wanted to date the sister of a Sergeant in our squadron my boss had advised against it. Mike's boss did not see a problem. Anyway we decided the Major's daughter might enjoy having another girl along so we came up with the idea of going on a picnic to Palo Duro Canyon.

Palo Duro is the second largest canyon in the country and it lies in the heart of the Texas Panhandle, only around 20 miles from Amarillo. In some spots it is 1,000 feet deep. Mike and I had been there before. It had hiking trails, horseback riding and even a chairlift to the top of the ridge, When we went to pick up the Major's daughter it became apparent that when I thought I lost ….I won big time. Mike and I were both shocked. Thankfully she looked nothing like her photo when she was 13. We had a great time. She was only going to be in town three days but on the chairlift we kissed and I set up dates for the next two nights (Sunday and Monday). I don't remember where we went but I saw her Sunday afternoon, that night and again Monday night. I hated to see her leave but before we said goodbye Monday night she said "You know we hit it off pretty well and I had a great time. If I had been living here we might have had a long lasting relationship. I'm really sorry to tell you this but the reason I came to see my dad was to tell him I am engaged and see if he will come give me away when I get married". The shock did not wear off for several weeks.

Mike continued to date The Senior Master Sergeant's daughter and many times we double dated …. me with a different new date most of the time. We were at his date's house one night and were talking to her dad. He asked us something and we both said "Yes Sir" at the same time. He laughed and said,"You do know that it is normally the other way around. NCO's say Sir to officers. We both responded, also at the same time saying when we are in your house to pick up your daughter we say Sir. He loved it and I soon found that he mentioned it to quite a few NCO's in our squadron. Mike and I talked about it later….Air Force protocol be damned. This guy was a Senior Master Sergeant with over twenty years in service. He had a Master's Degree and was the Head Instructor in the Missile School. We should be saying Sir to him even if he had no daughter. Our attitude probably was the reason Mike and I were well liked by the Senior NCO's.

LACK OF DATING OFTEN FROM LACK OF MONEY

Second Lieutenants back in the early sixties did not make much. The number two twenty two thirty comes to mind. Unless you had a few extras like a housing allowance, flight pay or hazardous duty pay your take home was $ 222.30 a month. Money went a long way back then but we had to pay for our own meals, extra uniforms that we had not been issued, dry cleaning and in my case a car payment. If you lived on base (which was required for Second Lieutenants) you got no housing allowance. You were stretched thin and it took extra money for dating. The answer at least for me was Moonlighting. I was always resourceful in college with getting jobs. Why should the Air Force be different.

With Softball and Football I had spent many a night after games with team members at the NCO Club. The sergeant who was Club Manager and I were acquainted. I stopped by one day and made him a suggestion. Why didn't I draw up a monthly calendar showing what all he had planned. It would advertise Lobster Night, Steak Night, Band Night, Entertainment Night or whatever happened to be the special occasion, food or drink coming up each night of the month. I had drawn up a sample showing art with each caption. He could print them on his mimeograph machine. Copy machines we have today were not around. I had made a sample on mimeograph paper. I told him he could have the calendars available for NCO's to pick up at the entrance and I bet it would pick up business when they had something special either to eat, drink or be entertained. He loved the idea and so I started doing the calendars and being paid. I no sooner had copies of the first one printed that the Officer's Club Manager asked me to start doing one for them also.

Two moonlighting jobs and I added a third. A Major who was a Sojourner asked me at one of our meetings what I studied in college. When I said Advertising Design. He said, "Was that like being in Art Class"? I said well yes, that I had classes that were strictly Ad Design courses but I also took watercolor, oil painting, sculpture, lettering and several others which were considered art. He said that his wife loved to paint but she was horrible at it. He said, "Why don't you offer Painting Classes to the Officer's Wives Club. I bet they would let you use a meeting room at the "O" Club". By the end of the next week I had 15 wives signed up for an eight week (an hour one night a week) course in watercolor and / or oils. The manager at the club not only provided me the room but let me advertise the class on the monthly calendar. He was paying me to advertise my own business. The 15 soon expanded because some wanted me to also teach drawing pencil sketches, so an extra hour was added before each class. I had more sign up for it than the painting class. The talent was surprising. I had one young Captain's wife who was a real artist. My drawing ability was no where near hers but she loved me giving her work compliments and she was my biggest promoter as far as recruiting more wives. She also invited me to their home to look at a self portrait she was doing for her husband but that may be a story I save for the movie.

By now you can see that I came up with a few ways to increase my income. I also found a way to save on meals. I volunteered for extra duty as a Mess Hall Inspector. We had several different mess halls on base because we had so many students. Each one had to be inspected once a month ….luckily not all on the same day. Every time I inspected one, the mess hall sergeant would ask me to stay for lunch. I learned early on ….let 'em know the day you are coming and their best meal would always be served.

TEMPORARY DUTY (TDY)

After over five years in the Air Force I am convinced that every and I mean EVERY Government Agency could cut their budget by 25% and it would not affect operations. When budgets are done everyone always asks for more than they need, expecting at least some cut and even so the budgets still rise around 10%. Must allow for inflation. When they have their budget they must spend it all or they think there will be a cut for the next one. The Air Force was no different. Toward the end of the budget period there were many (probably unnecessary) TDY Trips planned to use up funds. I played the game like everyone else. I had several Temporary Duty trips which were informative and I learned a good bit but they could have easily been eliminated.

One of those trips was from Amarillo AFB to Mather Air Force Base in Sacramento, California. Sergeant Summers, a friend, Lieutenant Jake Laeufer and I were sent TDY to Evaluate Collateral Training Programs at Mather. We had those also, so the intent was to see how they did theirs and see if we could use anything they did to improve ours. Sounded good on paper, so the trip was approved. We were allowed three days travel time each way. We could have gone by rail or bus or private vehicle. It was a long drive but Sergeant Summers had just bought a new car and wanted to drive. We would each get a mileage allowance which would defray costs by splitting the gas three ways. Also we would have to spend at least two nights on the road. It was considerably over a 24 hour drive (even 20 today with interstates) Summers had been in Japan during WWII and had married a Japanese woman. She still wore Kimono's and practiced the old tradition of walking 3 steps behind her husband. She had relatives in Sacramento so he planned on taking her along, as well as their pet dog. It was a male Jack Russell Terrier and liked only them (exclusively).

This meant Jake and I sat in the back seat. The trip started out fine and we were making good time. Summers had us leave before daylight because he wanted to put in a long drive our first day, We had to stop several times to walk the dog but we made it to the desert by mid afternoon. I believe we had been on Route 66 the whole way. Wherever we were, we found out the nearest town was 70 miles away. All at once the motor froze up and we were stopped on the shoulder of the road. His brand new car just stopped. His wife and the dog were not happy. Sergeant Summers was not happy (cussing up a storm). Jake and I were not happy and getting hot fast in the back seat. Of course the three guys all got out to look under the hood....like we would have the slightest idea what was wrong. After about half an hour a tractor truck driver stopped. He was the one who told us the nearest civilization was 70 miles West. Sergeant Summers said the truck driver was going to take him to find the nearest wrecker.

It was a long four hours before he returned with a wrecker. He was going to tow the car to the nearest car dealer. I cannot remember the make but think it may have been a Plymouth. Now keep in mind wreckers back then were not like you see today where the whole vehicle can be pulled up on a flatbed trailer. This was the old "Hoist up the front end and drag it" style. The driver said the sergeant, his wife and dog could squeeze in the cab with him but he did not know what to do with Jake and I. By now we were soaking wet with perspiration. I said "You sure as hell can't leave us here in the desert. We will ride in the car while you tow it". He tried to say it was against the law but I guess he realized we were desperate. Finally he said " Okay but you guys are going to have to lay down in the back seat so no one sees you". Try to picture two adult size men trying to ride 70 miles , tilted on the back of a wrecker, hiding from the law. To top it off it must have been 110 degrees. We were too big to both lay low enough in the back seat so I finally climbed over to the front. The sway of the front was worse than the back and I fought off getting car sick.

Somehow we made it. The dealer had someone drive the sergeant's wife and Jake and I to a motel which was only about four blocks from his dealership. Summers stayed to call the dealer where he bought the car and to try and figure out what to do. We got his wife and dog settled in their room. The motel people were not happy about the dog but after hearing our situation made an exception. Also I am sure they were considering calling the police when two soaking wet young guys showed up at the desk with an older Japanese woman wearing a Kimono and holding a mean dog, growling and showing his teeth. There is no way you can make this stuff up.

Jake and I shared a room and since we had all our stuff, we showered and changed clothes. By the time we were dressed Sergeant Summers knocked on our door. He had talked to the owner of the dealer in Amarillo. He had already left for the day but I am sure by the time Summers blessed out whoever he talked to he got the owners home phone number. He was still boiling when he was telling us what happened. They were going to take care of whatever was wrong and thought the local dealer could have us back on the road by noon the next day. He and his wife were ordering something delivered to their room for dinner then they were going to crash. He said he would see us sometime in the morning.

I don't remember the town we were in but I do remember we found out they had a Community College. Jake and I decided to hitch hike to the campus and try to find some action. We got a ride right away and the young man who picked us up was a student and dropped us off at the student union building. I guess we still looked enough like college students because we were able to check out ping pong paddles and a ball at the desk and no one asked for an ID. We played a game or two before a couple of girls walked up. After a brief conversation about who we really were and then we told them what we had been through that day we asked if we could buy them dinner. They accepted. One of them had a car. We went to one of the college hang outs had a good meal, enjoyed their company and they dropped us back at the motel. When Summers asked what we did the next morning “We said, “Oh we had dates with two lovely girls for dinner”. I’m not sure he believed us. It took them longer to fix the car but we got back on the road by one p.m.

Maybe we spent another night but I actually think we drove straight through because I recall Sergeant Summers dropping us off at the VOQ and he and his wife going to her relatives and it was late. The next day we reported in uniform to the people we had set up our meetings with. Summers had more to do than we did. The officers Jake and I worked with had printed up some information for us to take back about their Collateral Training Program so basically we were through by just after lunch. Summers had wanted to head back to Amarillo around that time the next day. Jake said “What do you want to do”? I asked him if he had ever been to San Francisco and he said no. I said, “Me neither I think it’s only a couple hours away why don’t we go”.

That’s how our adventure started. We hustled back to the VOQ, changed clothes, took another change of clothes and our shaving kits just in case we decided to spend the night. Neither of had anything small to pack the few things we took so we each borrowed paper bags from a clean up cart. No idea how to get there but we were on our way. At the main gate we asked an airman. He suggested hitch hiking to downtown Sacramento and taking a bus. He said they run every hour so that's what we did.

The bus took almost two hours. When we got there we had no idea where we could stay. Thinking hotels would be way too high we found the YMCA. We checked in and left our sacks. First on the agenda ride a Cable Car. We did … all the way to Fisherman's Wharf. On the way back we stopped somewhere near Chinatown. We decided it was time to celebrate and have a drink. Turned out the bar we went into was owned by an ex Air Force Pilot. We started talking and as soon as he found out we were Air Force Officers we became his best buddies. We asked him what we should do with only one night in San Francisco. He asked if we liked baseball. We both answered yes, He said "The Giants and the Yankees are playing at Candlestick Park tonight, why don't you guys go". Sounded like a good idea. He told us how to catch a bus right there in Chinatown which would take us direct to the stadium.

No one told us how cold Candlestick Park Stadium could be in the month of May. We happened to catch the game that went 12 innings. We froze to death. Also no one told us about the wind blowing off the bay. At the end of nine innings the score was tied 4 to 4. We went and each bought a sweatshirt but were still cold. We hung on until the bottom of the 11th but decided we had to go get warm. We were on the bus headed back to Chinatown listening to the portable radio of an old Chinese man. The San Francisco Giants finally beat the Los Angeles Dodgers 5 to 4. We got off the bus and went back to the same bar. The owner bought us drinks …. in fact he bought us many drinks. He locked up around 2 a.m and we set around drinking and listening to him tell war stories until the sun started coming up. We left, walked back to the YMCA picked up our sacks and caught the early bus back to Sacramento. Never spent the night at the "Y". I think the Sergeant thought our story was as true as the one we told about having dates two nights before. No, you just can't make this stuff up.

The trip back was uneventful. We gave a good report of our TDY trip and in fact adopted some of the things Mather did with their Collateral Training Program which were better than the ones we used.

A "MAJOR" BLUNDER

Please note that I am not writing about things in chronological order but trying to include everything which occurred during my eighteen month tour at Amarillo. I received a call from Lieutenant Colonel Parker. He had been a great person to report to because I rarely heard from him. He knew I had been running the squadron before Major Stephens came in and doing the same after he died. He said "Well your ninety days are almost up. I know you can keep running the squadron as Commander but you know that by regulation it is time I send another Major to take command". I told him that I understood and looked forward to welcoming a new boss. He laughed and said "Yeah Capps, I can just hear the excitement in your voice". He and I used to cut up a lot. I never bothered him unless we had a real problem and he knew either Sergeant Summers or I could usually handle most things which came up.

The Major he assigned was William P. Cameron (not his real last name, but he was William P. and the last name started with a "C" same initials as mine W.P.C.). This would prove interesting on the day he arrived. I introduced him to our staff. While in the orderly room he noticed several things I had approved with just my initials WPC. He said, "Looks like we won't have to change much cause I will still be approving things like Lieutenant Capps". He only stayed about thirty minutes, went in checked out his office, chatted a few minutes with Sergeant Summers and left. I came out of my office and asked where he had gone. Summers was never shy about giving his opinion. He said "Gone for the day. He talked to me like I was one of the Airmen. I don't think I'm gonna like the son of a bitch". I believe I said something like you haven't given him a chance he wasn't here that long.

He did not stay long the rest of the week. He would come in, close his door, get on the phone, stay an hour at the most and say something like see you in the morning. We found this rather peculiar. He was the direct opposite of the way Major Stephens came in. Instead of wanting to oversee everything, he wanted nothing to do. When we had orders for him to sign, he would drop them on my desk and say just sign these above my name and write the word FOR. So I was signing everything William P. Capps For William P. Cameron. Worked for me. Summers and I were still doing business as usual. Some days the Major never came in at all. One day Summers came in my office, closed the door (which he never did) and said "Lieutenant, we got to talk". He said "We've got a problem and it's the Major. I did not tell you but last week when he came in at ten, you were at the JAG office, I tried to take him a cup of coffee, I smelled alcohol on his breath. I think the guy is an alcoholic. Keep in mind that with our Airman Briggs, Sergeant Summers had spent a great amount of time dealing with an alcoholic. I wasn't sure exactly what we should do. The first Monday which rolled around for the Commander of Tech School's Staff meeting I thought it might cause a problem. When it came time to leave, for the meeting the Major was a no show. Summers said that I had better go attend and just say that you think the major is under the weather. Under the table would probably have been more appropriate. I went to the meeting and since I had always been the one attending no one paid much attention. I did not have to lie about why I was there instead of the new Commander.

I say no one noticed but one did. Lieutenant Colonel Angus asked me to stay after the meeting. We walked down to his office and I told him what we thought might be going on. He advised me to just keep running the squadron saying “These things generally work themselves out”. By the next week they did. The Staff Meeting was postponed until after lunch because the Commander was tied up with the General. Lieutenant Colonel Parker had called twice to speak to Major Cameron. Of course he was not in. Then we got an unexpected call from Major Pantell. I knew him pretty well and loved talking to him. He was the Commander of the Foreign Student Squadron. He had students from all over the world taking classes, many from instructors in my squadron. His office was in the quadrangle next to ours and those four buildings were the mirror image of ours. It seems as though Major Cameron had stumbled into his office thinking it was his. Pantell’s Orderly Room and office was set up exactly like ours. Pantell told me that Cameron was drunk and he put him to bed in an empty room he had in his barracks. He said “Let me know what you want me to do. I think he is sleeping it off". I told Summers the new development. About that time Parker called again. Summers answered and asked Lieutenant Colonel Parker if he could hold a second. With his hand over the phone he "Said Lieutenant either you tell him or I will. We are not covering for that prick any more". I took the phone and brought Parker up to speed. He did not hesitate. He said, "Cut orders assuming command immediately and meet me in Pantell's office".

When I got to the office Parker was already there. He said,"We will wait here until the Air Police arrive. Then we will wake sleeping beauty up". I am putting him under House Arrest and confining him to his quarters. He lived in base housing but I did not know if he was there alone or with his wife. I knew he was married. That was an awkward moment if there has ever been one. Parker had me go with him and the AP's to wake up the Major. He said "You are under house arrest and hereby relieved of your command. Lieutenant Capps has assumed command of the squadron". He had sobered up enough to stare daggers at me. I never knew exactly what happened but he was gone longer than a couple of months. My JAG buddies said they thought he was sent to the hospital in San Antonio. I told them that if so, I hoped they would have better luck with him than they did with my Airman. Things were never dull and you have to wonder how some decisions in the

Air Force are made. I got the word I would again be relieved as Commander. You can imagine the look on my face and everyone in the Orderly room when in walks Major Cameron. Maybe they thought he was cured. I cannot say he was now a member of AA but he did seem different. He was cheerful and showed up regularly on time for work. Basically nothing changed on our day to day running of the squadron. Cameron attended the Staff Meetings but did not do much more. He did do one thing which was important. My ER (Efficiency Report) was due and he wrote it. Lucky for me our clerk who typed it up kept a copy and gave it to Sergeant Summers.

If you want to ruin an Officer's career or anyone else for that matter, you don't give them a real bad ER. You give them a mediocre one. Rate them right down the middle. A bad one can be thrown out by claiming prejudice by the one writing it but one rated down the middle was a killer. Cameron had rated me mediocre. Sergeant Summers took it straight to Lieutenant Colonel Parker. Summers told him that he could not let this happen ….and he didn't. Parker rewrote it himself. He knew the job I had done and was still doing. Within another two weeks Cameron was replaced by another Major named Gohagan. He was sharp. Interestingly enough I got to assume command of the squadron from him one more time before I left Amarillo while he was on a temporary absence. That meant that as a Second Lieutenant I had been the Acting Commander of one of the largest permanent party squadrons four different times, serving close to a year.

A CONGRESSIONAL OF MY OWN

Around the middle of the year I got to thinking about how the Air Force could use what I had studied. How Advertising Design could be put to use. Back then none of our military branches paid for advertising on radio, television or newspapers. It was all public service donations as far as the media was concerned. An NCO in our squadron had pulled a tour as an Air Force Recruiter. He had seen what I was doing each month with the NCO Club calendar. He said, "Lieutenant you should be in Recruiting". Having all those Air Force Regulations in my office occasionally came in handy. The Major had told me "Read 'em when you need 'em". I read up on applying for duty with the Recruiting Service. The first thing that jumped out was the requirement for an officer that they be a Captain or above. I thought this was pretty stupid. Here I had been filling a Major's slot as a Second Lieutenant, why should it take a Captain or above to handle a job in Recruiting?

I wrote my Congressman. There was a Congressman, George Huddleston who often came to my grandfather's annual watermelon cutting and

I had met him several times. I wrote to ask for his help in finding out how I could apply for that kind of assignment and let the Air Force make use of my degree in Advertising Design. I think I also told him I had served as Acting Squadron Commander as a Second Lieutenant for one of the largest permanent party squadron's on base. I included the application pointing out the requirement to be a Captain.

Coming from a U.S. Congressman it went thru the chain of command rather rapidly. In fact it made it all the way to the desk of the Two Star General who was the Commander of the Recruiting Service. I found out a year later when I met him that he said, "This young man may have a point". It could not come until I served my full tour at Amarillo but I knew in June of 1964 that my next assignment would be in September as the first Second Lieutenant to ever be assigned to Recruiting. The General said that they should give me a chance. Two more Lieutenants from Amarillo (including Jake Laeufer) copied my application and eventually joined me at my new post. I was a Second Lieutenant when I reported on the First of September but I was promoted to First Lieutenant on the Twentieth.

STILL AT AMARILLO AND MORE EXTRA DUTIES

I doubt I can remember all of the Extra Duties I had while stationed at Amarillo. They were so numerous. One of them called for me to inspect MMPNC. That stood for Medical Material Program For Nuclear Casualties. They were stored in large sealed kits at various shelters around the base. I hated the duty and it was the one I most often swapped with another Lieutenant and took their Town Patrol. The reason it was so bad is you went all over the base and it took all day. If the kit (which measured around 3 foot wide and 5 foot tall) had been tampered with you had loads of paperwork to complete. Many people knew that there was Morphine in the kit and some vandalism would occasionally occur. Keep in mind Nuclear attacks were still on the minds of everyone in Washington. It had not been that long since the Bay Of Pigs (April 17 thru April 19, 1961 plus also we had the Cuban Missile Crisis, October 16 thru October 28, 1962).

ARMED FOR ONE EXTRA DUTY

I along with three other Lieutenants were assigned extra duty as Paying Agents for students. We reported each payday to the Base Finance Office and picked up cash to pay students. If they were paid by check it would have been too many to handle cashing on base. Since we were carrying large amounts of cash (nothing bigger than a twenty) , we were authorized to carry a weapon. When I went to the Provost Marshal Office to check out my gun I was issued a 45 and holster. Now each Lieutenant had to qualify once a year with a 38 and I had done so in OTS. I had earned my Marksman Certificate. Wanting to improve, while at Amarillo I had gone to the Firing Range and picked up both a Sharpshooter and Expert Certificate. I could not hit the broadside of a barn with a 45. My hands were too small and it kicked like a mule. I asked why I could not have a 38. He said "We don't have holsters for 38's. I asked "What if I have my own holster"? He said, "Well bring it next time and I will issue you a 38". Of course I didn't have a holster but before the next payday I went downtown and bought one.

When I showed up at the Finance Office the other three Lieutenants pitched a fit. "Why do you get to have a 38 and we get stuck with 45's"? They did look terrible compared to the cool looking 38 on my side. I told the guys that maybe I could help them. I went back to the dealer next time I was in town and asked if I sold three more holsters would he give me back enough for mine to be Free. It was a done deal. The other three Lieutenants were happy and so was I. It was a huge responsibility. Amarillo was an open base at that time. Anyone could come through the Main Gate just by saying they were visiting someone. We were walking across the base sometimes with twenty or more thousand dollars in a satchel. I eventually had a NCO who taught in Aircraft Maintenance make me an aluminum case with a hasp and lock. We were given a printout showing the airman's name and rank and how much he was to be paid. If we came up short we had to pay it out of our own pocket, so we were real careful counting them out their money. We had to show up early at the Finance Office and count to be sure we started out with the total amount shown at the bottom of our list. They did occasionally make mistakes. An Airman First Class carrying a carbine would accompany each of us. One day I noticed that the carbine had no clip. I asked the Provost Marshal "What is he supposed to do if someone tries to rob me, hit them with the gun"? He just chuckled and said, "That is why your gun is loaded Lieutenant".

A really funny thing occurred one day when I was paying a group of Air National Guard students who were going through a class in the Tech School. I would sit at a table with the money. They were supposed to line up in alphabetical order and I would mark them off the list as I paid them. They would walk up to the table say for example "Airman Frank Jones reporting for pay Sir". I would rarely look up. I was busy counting out the correct amount of money which matched the name on the list. The Airman carrying the rifle was supposed to look at their ID while they were in line. One of the Guard members stopped in front of my desk and did not say a word. He just stood there. Finally I said "Airman are you going to report"? He said with a slow Southern drawl "Sir I don't want to report to a dumb Lieutenant I went to high school with". I looked up and sure enough it was one of my classmates. I said in a firm voice "Airman stand over against that wall, I'll deal with you when I get through". The others in line thought he was in trouble. When I finished I gave him his pay and asked which squadron he was in. Turned out it was the Squadron Commander, who I knew that had graduated from the University Of Alabama. A phone call and later the three of us sneaked off to town for drinks.

Most orders notifying you of an Extra Duty are pretty clear as to what your assignment is, however, I received one set which named four Majors, one Captain and me to serve on a 35-66 Board. One of the staff from the JAG office was also named so I went there to find out what my duties were and what exactly a 35-66 Board was. It dealt with homosexuality. Back in 1963 the military attitude toward homosexuals was totally different than it is today. This was also long before "Don't ask don't tell". If someone was accused of being gay they came before a Board of Officers. If found guilty by the Board they could recommend the person be discharged. The Majors and the

Captain were voting members. I was listed as a non voting, Recorder. I thought by that title it meant I would record or take minutes of the meeting.

The JAG Officer explained that No it meant that I would represent the Air Force and present the evidence to the board. If it were a court of law, I would be like the Prosecuting Attorney The accused would be represented by a real attorney. They thought this would make it more fair to the accused. They said that I had to study the evidence and present it to the best of my ability to the Board. If there were witnesses I could cross examine them the same as the attorney. They gave me access to reports from OSI (Office of Special Investigation) Officers from another base where the accused had been stationed . The write up was pretty bad and detailed. I was surprised that his old base had not handled the case there.

Instead they just transferred the Airman to Amarillo where it appeared he did the same thing he was investigated for at his previous assignment. The report showed two witnesses had testified under oath that the accused had molested them in the barracks and forced them into committing homosexual acts. My first question to the lawyers in the JAG office was "What if two guys just didn't like the guy and they made up similar stories to get him kicked out of service". They said well that's probably why he was just transferred. If it happened again at this base it would add validity to what they said happened. I had asked that question before they let me see what he had been charged with at Amarillo.

When they let me read the Amarillo report it was even worse. Then they sprung the news that I could present this report to the board but I was not allowed to use anything from the OSI report from the previous base. They then told me their attitude toward homosexuality. They said we know this lifestyle exists and even in the military, but as long as it is a private thing we tend to ignore this regulation.

Out of the closet was not that common in 1963. The Air Force was big on initials. You've already seen many in this book (OTS, ATC, TDY, JAG, MMPNC). When I had to start learning all of them, I told my parents (Pun Intended) it was the Initial Phase of my Air Force career. Just think how excited the Air Force would have been with LBGQT if it had existed back in '63.

The JAG Officers then asked me how I felt. I told them that I had been raised to respect race, religion and lifestyle. My parents always taught me to treat all people how you want to be treated. Besides, since I majored in art I knew many men who were probably into that lifestyle. Many of them were talented artists and we treated each other with respect. What they did in their private life was none of my business and what I did was none of theirs.

Reading both reports I could tell this kid was bad news and really did not belong in service so I said I will try and present the case and not use anything from the OSI report.

If counselors were available back then to service members I was not aware of them. The closest thing we had was the Base Psychiatrist. I think somewhere in this book I mentioned he was my neighbor in the BOQ. He lived directly across the hall. I personally thought he was a "Nut Case". I would not dream of sending anyone in my squadron to him for counseling. You see the air and heating vents from my room (high on the wall) connected across to his room. I could hear him talking to his TV. If Superman was on and someone was about to be stabbed with a knife, he would scream "No No not with the knife" Yes, I thought he needed to see someone in his own profession. Of course I am sure he could also hear what went on in my room. If I had a guest I would often point to the vent and say shhhhhh.

The accused Airman's Squadron Commander had sent him to see the good Doctor; the Base Psychiatrist who was a Lieutenant Colonel. I had presented everything I was allowed to show the Board. It was tough not being able to present the other more detailed investigation which I had read in the OSI report. And believe me, it was detailed. The lawyer presented his case and called the Doctor as his witness. He got up and went through a list of questions he had asked the accused. His main point which the lawyer had him repeat several times was that in his interview the Airman had talked about his relations with the opposite sex and how he enjoyed having sex with his girlfriend back home. I asked the Major, who was President of the Board if I could cross examine the witness.

I said “Colonel, when you interviewed the Airman was he aware that these charges were being presented against him and was he also aware he would be coming before this Board”. He answered “Well yes but”. I interrupted and asked “Do you think he is intelligent enough to make up stories about relations with women when talking to you if it will help his case today”? He said “Yes I think”. I interrupted again and said “No further questions Doctor, thank you”. And I sat down. The lawyer tried to ask the same questions they had gone over before but I could see the looks on the Board member faces that they had already made up their minds. They voted to kick him out with a court marshal on the 35-66 Air Force Regulation.

I heard later from my JAG friends that when the Colonel who headed up the Judge Advocate General's Office saw the verdict, he rescinded it and granted the Airman a medical discharge. I knew the Colonel quite well from Sojourner's. A few nights later in the Officer's Club with his wife and friends, he called me over to the table and told the story of how he had to overturn a Board Ruling. He said "Clarence Darrow here did too good a job …. I want to get him transferred to my staff".

EXTRA DUTIES WHICH DID NOT COME OFTEN

One of those I only received orders for three times while stationed in Amarillo. It was School Staff Duty Officer. This was different from being Officer of the Day for the whole base (which I only did twice). For it, you stayed in the Base Command Post. This was Duty Officer for a day at Tech School and you did not stay in one place.

The orders spelled out your duties such as: Review the noon troop formation. You are to insure that troops in formation maintain proper military bearing, alignment, interval, distance, cadence and demeanor. Receive orders from the Commandant of Troops (that was my boss Lieutenant Colonel Parker). Contact the School Disaster Control Officer for access to the "Emergency Action File" which you are required to read. My old buddy Lieutenant Colonel Hobart was the Disaster Control Officer. You might know he would have that title. Junior Officers thought he created many disasters of his own.

Notify the Base Officer of the Day of your location at all times and how you can be reached. You may be called upon to assist the Foreign Students Branch and greet new Foreign Student Officers. That was my friend Major Pantell, and he did call.

Should a Broken Arrow Alert be called during your tour of duty you will be required to represent the Technical School at the Center Command Post. If this situation arises, your duty will be to observe and assist the Base Disaster Control Officer as he may direct. You hoped that did not occur while you were on duty (especially with Hobart the officer).

NEAR MISS ON A BIG ALERT

Above I mention I pulled duty as Base Officer of the Day only twice. You read about one time when I was on duty and helped the Colonel from the Inspector General's office with his drunken situation. My other tour in this job was a bit more exciting. You see I was Base Officer of the Day on November 21, 1963. I was coming off duty and turning over command to a Captain the morning of November 22 the day President Kennedy was shot.

When I say all hell broke loose it was an understatement. No one knew exactly what happened. I was still there when word came in I guess it was on a teletype or some type of printer back then. Of course the phones started ringing everywhere. Since no one knew if this was the start of an attack, the base immediately went on High Alert. I was told to get back to my squadron and set up our Command Post. I still don't know why we even had such a thing. If we were being attacked I am not sure what an Instructor Squadron would have done …. throw books and manuals at the enemy? My first thought was can I go to the Provost Marshal and get the 38 revolver I got issued each time I was the Pay Officer. The Real Air Force which we shared the base with, the SAC (Strategic Air Command) had all of their B 52's in the air within minutes. It was a sight, watching all those large planes take off one after another. Amarillo was always windy and with the wingspan of a B 52 being 185 feet, seeing them take off was like watching birds with flapping wings.

I was back in the Orderly Room (now our Command Post) when word came of the Presidents death. I still have a copy of the message from Secretary Eugene Zuckert and General Curtis LeMay. It read, in all caps:

PERSONAL FROM ZUCKERT AND LEMAY. ONE OF THE GREAT TRAGEDIES OF ALL TIME HAS TODAY TAKEN THE LIFE OF OUR PRESIDENT AND COMMANDER IN CHIEF. SINCE 20 JANUARY 1961, HIS LEADERSHIP HAS STRENGTHENED OUR FORCES AND BLENDED TOGETHER THE MIGHT OF THE FREE WORLD. OUR NATION AND OUR ALLIES WILL EVER BE GRATEFUL FOR THE STRONG HAND WITH WHICH HE SO MAGNIFICENTLY STEERED THE COURSE OF FREEDOM AND PEACE. AS MEMBERS OF THE ARMED SERVICES YOU ARE TRAINED TO CARRY ON DURING TIMES OF STRESS AND GRIEF. WE ARE CONFIDENT YOU WILL DO SO NOW, IN THE TRADITION WHICH HAS MADE OUR NATION GREAT. TO PRESIDENT KENNEDY'S FAMILY, TO THE NATION, AND TO THE FREE WORLD, WE, ON BEHALF OF THE MEN AND WOMEN OF THE UNITED STATES AIR FORCE, HAVE EXTENDED OUR DEEPEST SYMPATHY. IN THE KEEPING WITH OUR NATIONS GRIEF, THE NATIONAL FLAG WILL BE FLOWN AT HALF STAFF UNTIL SUNDOWN, 22 DECEMBER 1963.

We like everyone else stayed glued to the television for several days. Even though it did not appear an attack had anything to do with the assassination the base remained on alert status for quite awhile. The Air Training side of the base actually stayed on alert long after the SAC Wing went off. Somehow our side was late getting the word. I still have the photo of President Kennedy which I was told to remove from the Headquarters wall and replace it with a photo of Lyndon Johnson. Being stationed in Texas on that day was truly an experience. It was almost like the whole state felt responsible.

THE EXTRA DUTY WHICH WAS NOTICED

One day Colonel Parker called and said "Didn't you tell me your degree was in Advertising"? I said, "Well it is in Advertising Design". He said, "That's what I thought. I just volunteered you for extra duty in Base Beautification". I told him that I was already the Beautification Officer for our squadron. He said, "No they want you to try and get some publicity for the Base Beautification Program". That sounded like something I would enjoy doing so I asked "Okay where do they want me to start"?

Parker said "You do know that this is the pet project of General Monteith". He was the Major General who was Commander of the Technical Training Center. Parker said, "You need to report to Major Brown in Base Ops (I wont use the real color and you will see why). The name Major Brown rang a bell. The scuttlebutt around the "O"

Club that a Major Brown had been caught having sex with the wife of an NCO when the Major was sending the Sergeant on TDY trips. Everyone assumed it was a Major Brown who was a SAC pilot who was known to flirt with all the ladies at the club. When I walked into the office of the Major Brown I was to see, he would have been the last one who you would have suspected. He was tall. Thin and not what anyone would call handsome. He said “Are you Capps”? I said “Yes Sir, I was told you wanted me to help with some publicity on the Base Beautification Program”. He laughed and said “Help my ass … no I’m dumping this shit on you”. With that he shoved a big folder across his desk. He said “You may have heard but I am sorta in some trouble and they are reassigning me to another base. You better do a good job on this or the General will have you shipped out too”. I left his office wondering … does this now mean I am the Base Beautification Officer ...for the whole base. My duties at the squadron for beautification were simple. If the grass needed cutting I asked Sergeant Summers to get someone on his Duty Roster to cut it. Other than have one of our airmen in the office plant a few flowers at the entrance to the building that was it.

I stopped by Parker s office on the way back and told him of the exchange. He thought it was funny and said “Don’t worry I bet you can do a good job”. By now I had the reputation of being what was called a “Dog Robber”. That was a term which in the old days meant someone (usually an Airman) who ran errands for an officer. In 1964 a movie starring James Garner brought new meaning to the term. He was an officer who got things done and acquired items but not by the book. I had often gone around regulations if I thought there was a better way to solve a problem. I always remembered what Major Angus said “Regulations were written for people who couldn’t think for themselves … read ‘em when you need ‘em”. I don’t guess I was what you would call a “committed to a career” officer. I did not pay much attention to protocol because I knew I was to get out after my four year commitment. My ID Card had a date of separation (you will see why that was important later).

I was really fortunate to have been in service at the time I was. It had not been that long since the end of WWII and Korea. Many of the NCO’s and Officers in my squadron were veterans of one or both wars. They had stories to tell and I loved hearing them. I had gotten close to many of the Senior NCO’s. I called a group of them for a meeting (number around ten). I told them what had just been dumped in my lap. I explained about my experience in OTS when I devised a plan to get out of drill. Mostly I told them how I got to be good friends with the drill instructor and his advice on how to get the job in the Instructor Squadron. Also about him telling me never to forget that NCO's ran the Air Force. I said "I need you guys to help me figure out how to get this job done since there is no regulation, no budget and nothing authorizing it in the first place but it is the General's pet project. and he has two stars

They responded almost immediately. One of them said "First thing you are going to need is something to plant. Another piped up and said "Yeah and somebody to plant whatever you get and don't forget they will need tools. One of them joined in and said "Tools come under XB3….. Lieutenant that stands for "Expendable Base Level". Another chimed in "That means all Student Squadrons get issued tools and they are not on an inventory". I was busy taking notes. I came away from the meeting fired up and feeling that they were all willing to help me steer around whatever obstacles I ran into trying do an unauthorized job with basically no money and no staff.

First thing I came up with was get something to plant. Amarillo Air Force Base looked about as barren as the desert. We had a few trees and grass but little else. I started calling all the local nurseries downtown. I asked to speak to the owners whenever possible. I explained my task and asked them if they ever had anything, bushes, plants , trees, flowers, which looked like they might not make it. If they looked too bad to sell would they be willing to donate them to the base. We could try to save them. Interestingly enough they all responded in the affirmative.

Looking like I might have something to plant, now I needed somebody to plant whatever I came up with. The first load I went and got in my car. With the top down I put quite a few shrubs, bushes and even some flowering plants. While I was at the nursery the owner gave me another idea. He said why don't you contact the City. They are running some new roads and they are going to tear down some old houses. He said "I bet they would let you have anything you are willing to dig up from any of those places. I called them (took awhile to talk to the right person) but when I told him what it was for he said sure. He told me to come by his office and he would write a letter authorizing me to have whatever I wanted. He even gave me a map with addresses highlighted.

Now I needed labor. I went to Colonel Parker and told him I had a great idea. Students were always arriving early for their classes. The ones that did were called PATS Students (Personnel Awaiting Training). Every student squadron had at least several. I told Parker that they would make great planters for my supply of stuff. He said well you cannot make them do something like that as it was not authorized unless it was on an order. I asked if the Squadron Commanders could cut orders for them to be on temporary assignment to me until their class started. He said No. I came back with "What if I got them to volunteer"? He said "You've got to be kidding. You expect young airmen to volunteer to do yard work"? I asked "On average how many PATS students are there at any one time"? He said, "Anywhere from 50 to a 100". I said, "Okay get all you have now to come to the Base Theater and let me talk to them. I will bet you a steak dinner I can convince at least 20 to volunteer". He could not pass up my bet.

I called several of my student squadron Commander buddies to get an idea how long some of their troops had been sitting in the barracks waiting for their classes to start. The average was at least a week....some even longer. Parker had them march to the Base Theater the next day. He came along as well as two other officers in his office. He was the Commandant Of Troops so some of the kids were a bit intimidated when they saw him there. I went up on stage. I introduced myself and to Parker's surprise I introduced him and both of the officers. I said I heard that most of you guys are bored sitting around all day in the Barracks doing nothing. I know occasionally you get to march and I know you love doing that (heard a few moans with that remark). Well I am here to offer you an opportunity to get outside and not to march. I want volunteers to help us beautify this base. We can even help beautify your squadron barracks. I am about to start a competition between all student squadrons to see which one is selected "Beautification Squadron Of The Month". I am sure your Squadron Commander would like to win that.

I actually came up with that idea as I was talking but I'm thinking it looks like I have their attention. I went on to say my volunteers will have the opportunity to get off base. We may even have a picnic in town. I'm thinking that I bet I can get some of the Mess Hall Sergeants (who I have inspected) to pack some sack lunches. I am making up stuff as I go along but I can't stop now. I've got a steak dinner at stake. (steak for a stake had a nice ring to it). I need 20 volunteers to plant trees, plants, shrubs and flowers for Major General Monteith's pet project the Base Beautification Program. The sign -up sheet is by the back door. Out of around 70 in attendance 54 signed up. I picked the first 20. Parker and his two assistants were just smiling and shaking their heads. The steak by the way ....was great. Now I have troops. I need tools. I remembered what one of my NCO's said XB3 (expendable base level). I called each Student Squadron Commander and gave a list to each one of what I needed. I got a shovel from each, rakes hoses and several other tools like saws and clippers. Piled in the rear seat of my car ….I need a truck. I called the Motor Pool. It was run by a civilian …. a GS something (about equivalent to a Major). He said “No way...if it is not authorized, I cannot do it”. I called some of my NCO’s. They were now my unofficial Board Of Directors. I told them what I had run into. One of them had an idea. He said “Let me make a couple of calls”. The next day he came by my office and said “I know you are going to think this is crazy but can you get ten cases of toilet paper”? I said “Sure that would be easy. I can get one from each student squadron. If tools are expendable I know dammed well toilet paper has to be”. He said that a buddy of his was an Army Sergeant with the Texas National Guard. They cannot get enough toilet paper but they have vehicles which are never used. He wants to trade ten cases of TP for use of one of his weapon carriers. He will sign it out to you on a hand receipt.

Now that is how Lieutenant Mike Rex and I happened to be driving out the Main Gate of Amarillo Air Force Base with the top down on a Comet convertible with ten cases of toilet paper stacked and tied down in the back seat plus a couple in the trunk for good measure. We were in uniform so the guards at the gate saluted but had this “What the hell is this” look, on their faces. I had no clue how to drive an Army Weapons Carrier with all those gears. Somehow I made it back to my office but I may have driven most of the way in low gear. I’ve got wheels, I have troops, I have plants. The first day I tried to supervise. We did not have much to plant. Only the things I had brought in my car. One of the airmen knew how to drive the vehicle but he did not have a military drivers license. I wasn't worried about him driving on base but I did not want to chance letting him drive downtown. I needed someone to take charge of my squad.

CAPPS' RETRAINING GROUP

I can't take credit for the best idea which came for my unofficial band of merry men (most were boys between the ages of 18 and 21). One of my NCO's was a Master Sergeant who lived in the Orderly Room barracks. Since I had been on the Promotion Board as one of my extra duties I had seen many of the records of personnel in my squadron. This particular sergeant had at one time been a Major. He had been RIFed. This was a Reduction In Force. At that time he somehow was able to stay in as a Master Sergeant until he had his twenty years and he would retire with a Major's retirement pay. I am not at all sure how that worked but he was a real help to me as a Master Sergeant. He said "Lieutenant have you heard about the Retraining Group on base"? I had no idea what he was talking about.

He said, "You have a good relationship with the JAG office and they pretty much control the Retraining Group. It is where they reassign anyone who is in trouble where it will take awhile for their problem to be solved. For instance if an NCO is charged with a crime off base in a civilian court it could take a long time to go thru the system. I know someone who is in that group who would be the perfect man to run your beatification squad. And he can drive your big truck.".

I got the details and headed straight to the JAG office. The guy he was talking about was a Master Sergeant who had been an instructor in the 3324th Instructor Squadron. It was much smaller than ours. The Sergeant had been living off base with a girlfriend. She was a divorcee with a small child. They broke up...she had wanted money. He refused and she called the police and had him arrested claiming he had tried to drown her child in the tub. It got messy. A civilian lawyer had gotten him out on bail and released back to the military. They could not leave him on as an instructor because it had made the local paper. Therefore he was just sitting doing nothing in the Retraining Group. The JAG officers told me they did not believe a word of what the woman had said and thought she was just out to get him because he refused to give her money. I asked if he volunteered to be in my little Beautification Group would they assign him to me. They encouraged me to go talk to him. They thought it was a great idea. I talked to him and was impressed right away. I told him what I needed him to do and he hardly let me finish before he said “ I want to do it …. as soon as possible”. He said, “ Lieutenant I will do you the best job ever. You will be saving my life”. I never mentioned why he was in the Retraining Group. Best job ever would be understating what he brought to my new rag tag team.

He jumped in the very next day and more than took charge. He set up a schedule for rotating the many volunteers we had from the PATS Students. Word was getting out about how much they enjoyed getting out of the barracks and more and more wanted to work in our group. He had two teams working every day. One stayed on base and planted at various locations and one went with him to town to dig up new things. The City had found several extra places for us to find items for replanting and the nurseries were adding things daily. My new man in charge took everything off of me which allowed me to concentrate on just making the contacts. I even started thinking about publicity which is what I originally thought was all I was supposed to be doing.

I found a Trophy Business in downtown Amarillo and stopped by one day to ask about the price of a trophy which I could rotate among the student squadrons to be awarded for “Beautification Squadron Of The Month”. I had come up with that idea while speaking to the PATS Students in the Base Theater. Some of the Commanders of those squadrons started asking me what happened with that idea. When I told the trophy guy about what we were doing and the fact we had no budget but I was going to try and buy a trophy with my own money he said, “I am sure I have enough spare parts from broken trophies that I can fix something you can award which I will donate to the cause”. He came up with a trophy about three foot tall which almost made the “Commander’s Sports Trophy” look cheap.

That competition took off like a rocket soon after the first winner. The student squadron’s loved to compete with each other and when that first squadron had that trophy on display for a month the others put on the gas. Each squadron actually had some petty cash called a Unit Fund. They all started using some of that money buying flowers, pretty stones and mulch to spruce up around each building. Since they were using their own squadron personnel to do the work we didn’t have to use my PATS crew. Of course my Sergeant, now running that detail on a rotating basis, was able to help some of the commanders use their PATS Students who now had experience plus he would give them a few shrubs and plants every so often. He became extremely popular with the Student Squadron Commanders. They all wanted that trophy.

I went back to trying to figure how to get us some publicity for the program. The young lady who I dated who had appeared in the paper (when I won the bet) was still dating the guy I introduced her to and she and I had remained friends. Her dad worked at the Amarillo Paper. I called her and asked if she thought her dad could help. She asked him and he told her to have me call him and explain what we were doing. He was nice on the phone and said, "If you could come up with something big to plant, like maybe a huge tree, something unusual, I bet we would run a photo and help you publicize what you are doing". I thanked him and said I would work on it and call him back. One of the nursery guys, who had donated more than any of the others, I talked to often. I called him to ask if he had any suggestions telling him what the newspaper had said. He said . "You know if you transplanted a large pine that would be news. They are usually hard to move and that would probably make a good photo". He went on to say "You told me about the City allowing you to have anything from the yards of those houses being torn down. Maybe one of those has a tree you can move". He said that if I found one he would be happy to go with me and make suggestions as to how we could transplant it. He said "Not all trees grow well in our Panhandle soil so moving one might be a trick".

At one of the houses where we had moved a bunch of plants my Sergeant found a beautiful pine tree. It was at least thirty-five feet high. When the nursery guy and I went to see it I think he said it was an Austrian pine. He said that it would require a lot of digging and we would need to put a big burlap ball around the roots. He said “I doubt you have the equipment you would need to move something this big”. It was beautiful and all I could think of was what a great photo op it would make. After we got back to the base the sergeant and I discussed it and he said, “ Lieutenant, so far you have found contacts to pull off everything we’ve done so far. I bet you can come up with something”. I called another meeting of my unofficial Board Of Directors. It was the first time my NCO who had been getting everything done met with them. Turned out he and most of them were already friends. I had Polaroid photos of our tree. The first thing one of the NCO’s said was “ You are going to need a tractor trailer truck. You sure can’t get that in the Weapons Carrier’. Another chimed in “Yeah and it will take a crane to get that sucker out of the ground”. I said “Well at least I have already solved the problem of wrapping the roots in burlap. I’ve got three mess halls saving potato sacks for me”. At least that got a laugh.

I am not sure how we got it done but it took several of the sergeants crews about a week to dig up (by hand) that tree. We got a tractor trailer from the SAC Wing and also the large truck with a crane. It was what they used to lift Hound Dog Missiles (which B52's carried). We balled up the roots in those burlap sacks and we moved that tree to Amarillo Air Force Base. The SAC guys sent over a bigger crane to lift the tree high and drop it in the big hole the sergeants crew had dug. He had around 25 guys there that day. The Amarillo Paper had a photographer and a reporter there. It was a big deal for a Saturday morning. I was so excited the next morning when the Sunday paper had a photo in full color on the front page with the cut line "Amarillo Air Force Base Beautification Program In Full Swing" The tree was swinging from the crane about to be lowered into the hole.

I was still beaming at my accomplishment at lunch at the Officer's Club when I was paged for a phone call. It was Colonel Parker. He said, "Capps have you seen this morning's paper". I said "Yes Sir wasn't that great publicity"? He paused a moment and then said, " I think you need to come to my office". I was stunned. I think I mumbled something about not being in uniform. He said, It doesn't matter just come now". I could not imagine what was wrong. Had something happened to my family and he didn't want to tell me on the phone. Anyway I hurried to his office. When I went in he was also in civilian clothes. He had a copy of the paper on his desk. I sat down in the chair opposite his desk. He said “You ever heard of General Thomas Fairley”? (not his real name) I said “No Sir”. He said, “ Well he happens to be the Commander of the Army National Guard for the State of Texas. He just called Major General Monteith and asked him if he had seen the morning paper. When the General said no, he said, You better look at it and tell me what the hell you are doing with one of my Weapons Carriers on your Air Force Base”. I had not noticed … maybe because I had gotten so used to seeing that truck every day but there it was on the front page in living color. The photo of the tree but also clearly in the center of the picture was the Weapons Carrier with printed on the door “ARMY NATIONAL GUARD”.

Parker said, “Knowing you. I am sure you have a logical explanation”. I stuttered but told him how I had tried to get a truck from the Motor Pool and when they turned me down I still needed a truck. I said, “I have the truck signed out to me on a Hand Receipt. I traded twelve cases of toilet paper for it”. Parker could not contain himself. He was laughing so hard. He finally calmed down enough to say “You need to call General Monteith right now. I wouldn’t miss this for the world ….when you tell him what you have done”. Colonel Parker actually dialed the General and talked to him first. He stayed on the extension so he could hear. He said, General I have Lieutenant Capps on the line.

I think he can explain what happened". I went through the whole story, maybe a little nervous, but I told it exactly how I told Parker.

The General was hysterical. He was laughing so hard. He finally said something to the effect that I was doing a fine job and being extremely resourceful while working under handicapped conditions. To make a long story short my rag tag team was responsible for planting over 3,000 trees, plants and shrubs on Amarillo Air Force Base. It sure looked better than when we started. For an effort which was not authorized, funded for ,or covered by any orders or regulation, I could not receive any official recognition. I did, however, eventually get an assignment that I feel was a direct result of what I had accomplished.

NOMINATED FOR ESCORT OFFICER

April first 1964 was on a Wednesday. Colonel Parker called to ask if I had looked in the "IN" box on my desk. I had not and he just said "I left you a list of countries. Look it over and pick three you would like to visit if you got the chance. Put them in order of preference. You may get a temporary duty assignment to one of them.". He then added "This is not an April Fool joke, just call me when you decide ….can't tell you any more". Out of 16 countries I picked Sweden, Switzerland and Germany.

On April 4th (happened to be my birthday) I received a copy of a letter signed by Major General Monteith. It was to a Colonel Asworth, Commander, Civil Air Patrol, Ellington Air Force Base, Texas. It read: 1. I have selected the following as Escort Officers for the International Air Cadet Exchange Program. It named First Lieutenant Richard Ahart and Second Lieutenant William Capps and gave our preferences. Number 2. said: You are most gracious to enable us to reward these two fine young officers who will, I am sure, represent the USAF in a commendable manner, We will await further instructions from you on responsibilities, reporting dates, etc.

I was in a state of shock but pretty excited about the possibility. It was not until April 23rd but on that date orders came in informing me I was selected as an Escort to take a group of 15 Air Cadets to Tel Aviv, Israel. First Lieutenant Ahart was to escort a group to Ankara, Turkey. Neither of us got countries we had chosen but it did not matter. On the same set of orders it listed all of the officers chosen and where each would be going. Of the 16 countries, the escorts consisted of 3 Lieutenant Colonels, 2 Majors, 7 Captains, 3 First Lieutenants and 1 Second Lieutenant (me).

It was a long Temporary Duty. With travel time it would run well over thirty days. The orders were not too specific but I had until the end of April and the months of May and June and part of July to get the details. Ahart and I would leave around the 14th of July and fly to Washington, D.C. There we were to receive a few days of briefings and meet the Cadets who would be going from the United States. I spent those next months studying Hebrew and trying to learn as much about Israel as I could. Trying to figure out how I got that country was a bit of a mystery. It was listed last on my set of orders but then I figured out they were listed by the rank of the officers who were the escorts and I was the only Second Lieutenant so my name and Israel came last. My suspicion was that someone saw that I was half Jewish because when I got to Israel I was told that since my mother was Jewish, I was Jewish. They accepted me as a "Lonsman" which was more Yiddish and basically meant kin or fellow Jew. The absolute worst part for both Lieutenant Ahart and I was our orders called for "immunizations per an Air Force regulation will start immediately". Because of the countries we were both being sent I think we had to have every shot which had been invented. Luckily they spread them out over several days. We also filled out forms for a passport. I would not find out that it was actually a Diplomatic Passport until an Emergency arose during the return trip. In Washington I met James Mitchell who would be the civilian Escort with me. He was a Colonel in the Civil Air Patrol. We decided that I and all our cadets would refer to him as Colonel Mitchell. He and I met our five cadets from the United States. They were from all over and I cannot remember from where although I know one was from California because he gave me the most trouble. When we got to Germany for another briefing at Rhein-Main Air Force Base outside Frankfort we met our other 10 cadets. We would be responsible for 15 in total. We had in addition to our five from the U.S., two from Canada, two from Denmark, two from Great Britain, two from France and two from Sweden. Some of the briefings were for Colonel Mitchell and I and some included all of our cadets.

Our cadets were told that they were not allowed to drink alcohol and that their escorts were in charge and had full authority over any cadet who misbehaved or brought any shame to the country they were representing. We could send them home. They put the fear in them by telling stories of previous Cadet Exchange trips where guys were sent back home for various reasons. The ages of our cadets ranged from 18 to 21. I was not much older since I was only 24.

The Colonel and I discussed it and decided that for our group we would make the alcohol rule flexible. We had two guys from France who were both 21 and they had grown up drinking wine. We told all of our cadets that as their escorts we were totally responsible and we were going to have our own rules. They could drink alcoholic beverages if they were served but only in moderation. We told them that if they had been chosen to represent their country we knew it was because they were responsible adults and we felt they could be trusted. If anyone got drunk they were going home.. We never once had a single problem. When we returned back to Germany after the trip for our debriefing we heard horror stories from one of the Lieutenant Colonels and a couple of the Captains. Colonel Mitchell and I had a great group.

It turned out that Israel was a super pick for our country to visit. To begin with because of the way the Air Force had negotiated transportation we got to see several countries along the way. We left Germany in a German General's Jet. Neither the pilot or the co-pilot spoke English. One of them could understand some French so the Colonel and I communicated through out two French cadets (but not real well). Our first stop was Nice, France. We were dropping off cadets going to other countries. Next we landed in Naples, Italy then on to Athens, Greece. It was there that I said I wish we had time to get off the plane and at least see a little bit of Athens. "They said no problem, they would have engine trouble for two hours but then they had to get back on schedule". The spokesperson from Greece asked what we would like to see. In only two hours it could not be much as it would have to be close. The other CAP escort and the Colonel talked and decided to let the guy make a suggestion or two and let the cadets vote. He came back and said “I gave them a choice of seeing the Acropolis or dropping by a topless beach, both are close”. Guess which one they voted on. The German pilots let me stay in the cockpit while they buzzed the Acropolis so I could say at least I saw it. I am sure the buzzing was illegal.

From Athens we flew to Incirlik Air Base in Turkey where the Germans turned their passengers over to the Israeli Air Force. We flew from there to Tel Aviv, Israel in a Nord Noratlas. It was built by the French and looked like a small version of the Air Force “Flying Boxcar” the C-119. When we arrived in Tel Aviv there was a storm. I was in the cockpit sitting in a jump seat between the pilot and co-pilot, We made two passes which had to be aborted. The second one scared me to death because we barely missed the top of a hanger. They had to radio that we needed to go to an alternate place to land. This messed up the huge welcoming committee which the Israelis had set up.

Wherever the alternate strip was, must have been about 45 minutes away from the original. A group of the dignitaries apparently jumped on a fire truck because it came out to the plane with siren blaring to meet our group. They kept apologizing because it was not the way they had hoped to greet our arrival. The Fire Chief had a uniform with many ribbons just like some of the military personnel who came to meet us. Apparently uniforms were a big deal in Israel back in 1964. The Fire Chief was included in many of the functions we attended during our visit to the country.

The first night we were all worn out and stayed in a hotel in Maccabiah Village. The next day we heard a presentation at the Maccabiah Hotel on the re-birth and development of the State of Israel. After lunch the cadets met their host families. All of our cadets stayed with different families in their homes. The Colonel and I stayed at the Ramat Aviv Hotel in Ramat Gan, just outside Tel Aviv. We were treated like royalty at the hotel. He and I did not have a regular room in the hotel We had a cabana house out next to the pool. There were only around four of those. The manager of the hotel was constantly checking to see if we needed anything. Other guests wondered exactly who we were. I had been told during the briefings both in Washington and in Germany that I could wear civilian clothes except for certain events. Of course I wore it coming and going because I was on military planes (even though they were with the German and Israeli Air Force). Also the first day, being the only Air Force Officer my orders said my first official act was to report to the Senior Air Force Commander or the Air Attache at the American

Embassy. I was to give them a copy of my orders. I had flexibility as the orders also said variations in itinerary were authorized.

Saturday, Colonel Mitchell and I had a day to relax at the hotel while the cadets did things with the families they were staying with. I went to the pool. Not long after I was in the pool a really attractive girl approached and asked who I was and why I was in Israel. She had seen us arrive and knew it was unusual for the hotel manager to be escorting guests to their room. She asked if we were famous or something. I laughed and told her we were definitely not celebrities, who I was and why we were there. We enjoyed each others company all afternoon and even had lunch together.

Sunday night was one of the occasions where I would be in uniform. We were attending a reception at our hotel out by the pool where the Colonel and I and all our cadets and the families they were staying with would meet the Israeli General who was Commander of their Air Force. His name was Ezer Weizman and he would go on to be elected President of Israel in 1993. We would also be meeting the Israeli Air Force Attache. All of the men I met at our Embassy would also be in attendance.

I really wanted the young lady I spent most of Saturday with to see me in my uniform. When I found out she was not a guest but just a local who often came and used the pool because her dad and the manager were close friends I asked for a date for Sunday night to attend the reception. I told her I would get a cab and come pick her up. She said, No I will drive my own car. Just tell me what time to be here and what I should wear. I said that I thought it would be sorta dressy because of all the dignitaries who would be there. When I went back to the room and told the Colonel that I had a date for the next night I think he was surprised.

Sunday night was rather formal. Many of the Israeli Officers who we met at our first briefing at the hotel in Maccabiah were in their Dress Uniforms. The General arrived. He was a tall and imposing figure. A table with many hor d'oeuvres was being set up. Before they were to be served a receiving line was being formed. The General and the Air Attache were at the end. My date had arrived and she looked stunning. I introduced her to Colonel Mitchell and we proceeded to go through the receiving line. I would introduce myself and then my date. Everything was moving rather nicely until we got to the Air Attache. When I went to introduce my date she butted in and said "Hi dad". I was shocked and obviously so was he. She had not told him where she was going. He shook my hand again and said " Lieutenant, you work pretty fast, I understand you've only been in

Israel two days".The General, who was next in line also got a big kick out of it.

She and I stayed in contact for most of my stay and her dad would have to come to my rescue on a situation which later on almost became an International incident. After the receiving line broke up everyone mingled and ate from a lavish display of finger food. The Colonel and I got to spend some time with General Weizman. My date was being entertained by several of our cadets. My stock went up with them because they all liked my date. The General asked several questions about U.S. Aircraft. Luckily Colonel Mitchell being an Officer in the Civil Air Patrol had most of the answers. It was ironic that I was the only U.S. Air Force Officer there and the only one among our delegation who was not already a pilot or in training to be one. Several of our cadets had already had solo flights. Since my date had come in her own car, after the reception broke up she drove me around to see some of the neighborhood.

FLEXIBLE SCHEDULE

We had a printed schedule of events but other than our meeting with the President Of Israel and the Mayors of Ramat Gan and the Mayor of Tel Aviv we deviated when necessary or other opportunities arose. We were constantly on the go. Most of the time it was by bus. The cadets were picked up at various points from the host families and the Colonel and I were always picked up last. This gave us some extra time in the mornings to either sleep in or do something around the hotel. Many days I took an early swim but one morning I walked down to the tennis court. A guy there was by himself and asked if I played. I told him about my embarrassment of playing in the base tournament in Amarillo. He said for me not to worry he just wanted someone to volley with him. He was obviously from the U.S. Turned out he was with Paramount Pictures. They were shooting the film "Judith" starring Sophia Loren and Peter Finch.

I mentioned this to one of the Israeli Officers who was accompanying us to meet the Mayor of Ramat Gan. He seemed real interested and said that he would check to see if they could schedule our group to visit the set. A couple days later we were on our way someplace and they announced we were going to the set of a movie being shot not far from where we were. Maybe we would get to meet Sophia Loren. The U.S. cadets all knew who she was but only a few of the international ones did.

We arrive at the set only to find out Ms. Loren had chosen that day to have a tantrum and not show up. Finch was not in a scene that day so he was not on set either. We got to watch a scene or two being shot and the Director was nice to explain how everything was being done but we only stayed around 45 minutes. When the bus pulled out one of our guides got on the speaker and said "Did everyone bring their swim suits"? All the cadets had. The Colonel and I had not. He went on to say we were taking a short break at a hotel where they had invited us to take a quick swim and it happened to be the hotel where Sophia Loren was staying. He said : Who knows she may come down to say hello".

I guess they called her room and she refused. Our gang could have cared less. They all went to the pool. One of our hosts that day was an Israeli Air Force Colonel. He and I had been sitting together on the bus. We had gotten acquainted at the General's reception. He had a brother living in the states and we enjoyed talking. Colonel Mitchell and I were just killing time in the lobby when that Colonel came over and said "I have a real treat for you guys. Come with me. We are going to have tea with David Ben-Gurion". He was the primary founder of the State of Israel and the first Prime Minister. In Israel the President's powers are largely ceremonial; the Prime Minister holds the executive power. Getting to meet him was indeed a treat. We sat and talked with him for almost an hour. Coincidentally he was staying at that hotel and the Colonel who was with us happened to be a personal friend. Later that same week this same

Colonel would introduce me to another well known person.

I might as well explain that one now as well. Mitchell and I were busy every day with tours but when evening came our cadets went back to their host families. Each day different members of the Israeli military would go along with us as hosts. At night they thought they needed to entertain the Colonel and I. We went club hopping for about five straight nights in a row, getting back sometimes around 1 or 2 a.m. The Israelis were tag teaming, so they were doing it every night. We were exhausted. That same Colonel who introduced us to Ben-Gurian said, " Why don't you guys come to my house tonight for dinner. You can relax. It's my son's bar mitz-vah and there will be some special guests I want you to meet. The guest list was a who's who of high ranking military officers. The Air Attache (who's daughter I dated was also there). The big surprise was when the door opened and in walks Moshe Dayan (patch over his eye and all). He was a National Hero in Israel. He had been the Commander of the Jerusalem front during the 1948 Arab-Israli War. At the time we met him he was Minister Of Agriculture. He was small in stature but when he came in he became the center of attention. I happened to be sitting directly across the table from him at dinner. He became a hero again during the Six Day War of 1967 when he became Defense

Minister. Some say he became to the World, the fighting symbol of the new State of Israel.

When we went to that movie set we did not get to see either star from the movie. Later one night Colonel Mitchell and I were being taken to a bar where the entertainer was a belly dancer who had recently appeared in a James Bond Movie "*From Russia With Love*". When we walked in, there was Peter Finch (the actor we missed at the set). He was up right at the stage trying to get the belly dancer to pose in a photo with him. He was drunk as a skunk. He tried to sit down, missed his chair and fell on the floor. I guess that is when he broke his camera. He started yelling, as did two guys with him....asking if anyone had a camera? I just happened to have my Polaroid. I walked up, tried to introduce myself, but I doubt he heard anything accept the fact I had a camera and could not only take his photo but develop it and hand it to him. Polaroid was still basically unheard of in Israel. He got the dancer to lean down and hug him and I snapped the photo. When it developed he said thanks and asked one of his lackeys to hand me a dollar. I just said "Thanks but tip the dancer".

We went by bus to Jerusalem. While there we met with the President. At that time Zalman Shazar was the President. At this writing there have only been nine Presidents Of Israel so counting General Weizman who became President in 1993 I can say

I met two out of nine. We left the office of the president and went to the University, There we got to go to a vault and see the Dead Sea Scrolls. Leaving the University we visited (at that time the tallest building in the city) the YMCA. From the top we had a view of the wall which divided the city. On that particular day King Hussein of Jordan was giving a speech on the other side of the wall. We could see the crowds. I am not sure what happened but something broke out during his speech and we were suddenly herded out to our bus. Apparently whatever the disturbance was, caused our hosts to worry about our safety. For the next four or five days we had an armed escort (a weapons carrier full of troops all carrying Uzi sub-machine guns).

One of the days I rode in the truck with the troops. The Israeli Lieutenant who commanded them and I struck up a friendship. He was my age and spoke English fluently. In fact most all young adults in Israel spoke English. Anyway having those troops along added a little excitement. One day at lunch I was sitting with them and realized one of the troops put his Uzi on the table and it was pointing at me. I turned the barrel so it was pointing at the wall. I did not mean to get the young man in trouble but the Lieutenant saw me turn the gun and he chewed out the young soldier.

Another day their truck was following the bus and they started honking for us to pull over. We were passing a herd (I guess you call them a herd) of wild camels. One of the troops ran down with a rope and brought one up and let anyone who wanted to, climb aboard for a photo op.

A couple of weeks after we no longer had our armed escort an Israeli Major came to me with some bad news. The Lieutenant who I had become friends with had been shot and killed while on a routine patrol near the Goland Heights. They said it was a sniper attack. Even when they did not have a war going on they lived with danger on a daily basis. I took time out from our busy schedule and went with that Major to the funeral.

The Israeli Colonel and I had really become close. He knew that I was upset by the Lieutenant being killed. One early morning he and I were walking downtown in Tel Aviv. I said "Colonel I am not sure how to express this or even how to explain it and it probably sounds corny but when we are just walking I almost feel the spirit of the people". He said, "Lieutenant it is not corny at all and I can tell you what it is. We have been pushed as far as we will be pushed. We will all die here". I have never forgotten that in all these years. I witnessed that attitude first hand when we spent some time with the men and women of their Army at their desert survival training site. We stayed overnight. Our cadets slept in sleeping bags along with all of their troops. The Colonel and I had a tent. Everyone both male and female serve at least two years in the military. At one of the training sites I saw an 18 year old young girl fire an Uzi. They had spray paint cans set up on a fence about 20 yards away. She sprayed them, not missing a one. Paint colors were flying all around. Of course I am sure she was the one always chosen to perform for visitors. They let me try it and I think I hit 3. During my long stay in Israel I got to visit many (if not all) of their military facilities.

My favorite by far was the Flying School. Colonel Mitchell got to fly with one of the instructors in a Mystere. It was a fighter-bomber built by the French. They introduced me to their top glider pilot instructor. He took me up in a two seat open cockpit glider. We took off being pulled down the runway by a cable attached to a weight dropped from a tower. As soon as we were airborne the pilot released the cable. We were up for almost an hour. At one point he told me to take the stick (it was a stick controlled aircraft). I immediately said "I'm not a pilot". He said "They told me you were a Lieutenant in the U.S. Air Force". I said "I am but not a pilot". I guess all their Air Force Officers were pilots. He said "Don't worry, it's easy just take hold of the stick but go back left, that's the border by that group of trees". I said "Will they shoot us down if we go over the border"? He smiled and said, “Sometimes”. It was a wonderful experience. All you could hear was the sound of the wind. He kept watching for birds circling. We would go to where they were and catch a thermal and climb to a higher altitude. When it came time to land we were going horizontal to the runway. He said “Hang on”, made a steep dive turned and brought it in as smooth as glass. As we glided to a stop, guys were running along both sides to grab the wings because it only had landing gear in the center. I was told that every pilot in the Israeli Air Force learns in a glider first. They want them to know how to glide back home with anything they are flying if they get in trouble.

As small as the State of Israel is I guess that would be possible. It is only about the size of the state of New Jersey (also the shape is similar). Jersey has 8,722 square miles and Israel has 8,630.We went from one end to the other. We drove to Haifa in the North and we flew to Eilat in the South. Part of the country has the Mediterranean Sea on the left and part has the Jordan River and Dead Sea on the right. They are in close proximity to Lebanon, Syria, Jordan and Egypt.

Mitchell flew in the jet. I flew in the Glider and while this was going on they were taking our cadets up two at a time in one of two different Cessna aircraft's. When our two guys from Sweden had their turn an accident occurred. On coming in for a landing the pilot was too low on approach and the landing gear hit the top of some irrigation pipes which were in a farm field just before the runway began. The plane nosed into the plowed field and came to a sudden stop. The only person hurt was the pilot. He suffered a bloody nose and a bruised ego. All of our cadets fearing the worst ran to where it came to rest at the far end of the airfield. The Swedish boys came out laughing and of course they were still wearing their blazers. Every cadet had brought along a sport coat, shirt and tie for our more formal meetings. The guys from Sweden wore theirs every day no matter where we were going or how hot it was. Believe me in July and August it can get Hot in Israel. By the time I got down to the "Crash Site" our cadets were all taking photos. The plane was sitting up on it's nose. I did not realize that Israel had a strict rule to not photograph a military plane in an accident. A Colonel appeared out of nowhere. He had not been in our welcoming party. He proceeded to take up the film from all the cadets cameras and the one Polaroid shot which I took. The kids were having a fit. Many of the photos they had taken on our whole trip were on the rolls taken away by the Colonel. They wanted Colonel Mitchel and I to do something. Mitchell talked to the Israeli Colonel but he would not budge. I told everyone to let me work on it when we got back to the hotel.

I called the young lady who I dated when we first arrived. We had been out a few more times. I asked if her dad could help. She explained to him what had happened and he got on the line. He said maybe he could get them to have all the rolls developed and give us back those photos which did not show the plane. That's almost how an International Incident happened (which I thought was a bit trivial because who gives a hoot about a Cessna with maybe a bent propeller). A couple days later I picked up the stack of photos in the Air Attache's office. Deciding who took which photos since they all were similar was not easy but the guys worked it out. The only one who knew exactly which ones were his was one of the French cadets who had the only shots taken at the Greek Topless beach.

TEA AND COFFEE IN THE DESERT

One of our guides spoke fluent Arabic. As I said occasionally we broke away from our printed itinerary. He said "We are going to take the bus out in the desert. I know one of the Bedouin Chiefs and he has invited your delegation for a visit in his tent". Now we were in the middle of nowhere … desert as far as you could see in all directions. From what I observed no electricity, no running water. Home Garden Network could have done a feature called "Really Off The Grid". Outside the tent stood two camels, a pen with four or five goats, and of all things, a John Deere Tractor. The tent or should I say tents because it looked like three tents arranged together. The biggest was in the middle and I guess that served as the living or great room. We all sat around on large pillows on the floor which was covered by a big rug or Persian looking carpet.

Before we arrived our guide said we were not to look at any of his three wives. One, who looked from what you could see beyond their veils, to be much younger than the other two. She kept staring at me (I happened to be in uniform that day). I winked and she winked back. We were offered tea or coffee. When I saw that the spoon in the coffee stood up by itself I chose tea. The Chief spoke and our guide translated. The Chief was enjoying talking to such a large audience. Hell, he was probably so glad to see some other human beings. The cadets were getting restless and the Chief's speech was of no interest to anyone there. Our guide said "He wants us to stay and have dinner. He has offered to kill and barbecue one of his goats". I looked at my watch and said "Tell him we appreciate his hospitality but we are running late for our next appointment". Everyone was glad I said it. On the way back I asked the guide about the tractor because I did not see anything he had or could use it for. He said "It was a gift from all of his tribesmen and it was a status symbol. He did not have gas to run it anyway". You just can't make this stuff up.

When we went to Haifa (which is a major seaport) we (cadets too) stayed three nights at the Panorama Hotel up on Mt. Carmel. It has a great view of the city. From there we visited the port, Galilee, Caperunaum, Nazardeth and Acre. The last one, Acre was where part of the movie, "Exodus" was filmed. The manager had a photo of he and Paul Newman (who was in the movie) proudly displayed in his office. Even if Newman didn't stay there he may have at least had dinner in their dining room which was really good. It was leaving the hotel after our third night where our cadet from California caused us a problem. When we were all packed and on the bus he was missing. When I asked if anyone knew where he was two or three of them spoke up. They had all been bowling the night before. In the bowling alley our surfer boy had snagged a date with a local girl. He had caught a cab with her to, quote "see her home" after bowling. They thought she lived somewhere in the city. One of our hosts for the day said "We can give him thirty minutes but if he doesn't show we have to leave. We can tell the people in the hotel to just have him sit until we can send someone back to pick him up. He had about five minutes left when we saw him trudging up the mountain. He and the girl had stayed in town too long and cabs were no longer running. He had hiked all the way back up the mountain. He slept going back to Tel Aviv.

EILAT THE RIVERA OF ISRAEL

The trip from Tel Aviv to Eilat is only a little over a four hour drive but to save time we were flown in a military plane. I was supposed to wear my uniform but Colonel Mitchell talked me out of it. “He said who’s gonna care. Besides even some of our Israeli Air Force hosts were in civilian clothes and we were going to be going swimming down there. Eilat is the southernmost city. It is a busy port and a popular resort on the northern tip of the Red Sea. In Israel it is called the Gulf of Eilat. In Jordan they call it the Gulf of Aquaba. It is a huge tourist destination. Our cadets loved it. We flew back that night.

THE FIRE BRIGADE

When we arrived in Tel Aviv we had been met by the Fire Chief and an ensemble of people who could pile in and on the fire truck. That had become our welcoming committee because bad weather had forced us to an alternate landing site. I think I said at the time that the Fire Chief had a uniform with as many ribbons as most of the Senior Military Officers. Well the Chief and the whole Fire Brigade were important in Tel Aviv. The Chief had actually gone with our delegation to many of the functions we attended. Now it was his turn and his firemen who would become hosts. All of the cadets and quite a few of their host families (including their kids) were invited to the training center for the Fire Brigade. They put on demonstrations using real fire. They had about a six or eight story tall concrete tower where they demonstrated going up and down ladders carrying hoses and even bringing people out of it with smoke pouring from the windows.

They also had a new hook and ladder truck. They had extended the ladder almost straight up 100 feet. The Chief said we want to see the Lieutenant climb to the top. Of course all of the cadets started cheering and clapping and eventually chanting "Climb It Lefttenant". This had become my nickname because the two cadets from Great Britain had originally pronounced Lieutenant as LEFT TENANT. I was about to take up the challenge when the small cute 16 year old daughter of the Chief stepped forward and said, "Come on scaredy-cat I will show you how it is done". With that everyone cheered. She had obviously done it before because I could see the Chief smiling. They had set this all up. It had all been rehearsed. She was wearing a dress. She climbed up on the truck and announced "The secret to climbing up this high is you don't look down and since I am going first you might enjoy the view". Roars of laughter. I thought I had better go along and maybe add a little something to their attempt to embarrass me, so I said "In that case I better take along my Polaroid". We started climbing. When we got up about 20 feet (two stories), she paused and yelled "We have to rest up every now and then". With that she lifted her dress and threw it over my head. Again roars of laughter. I knew for a fact then it had been planned as she had on Bermuda shorts under her dress Still it was cozy under there.

The two of us climbed all the way to the top and took several Polaroid shots looking down. I did not realize how scary and how high up we were until I looked at those photos when we got down. The people below looked pretty small. Back on the ground my legs were shaking. They took a photo of us sitting at the bottom of the ladder with me with a fire helmet on and her kissing my cheek. I asked for a copy but never got it.

Before we left they announced that our farewell party, which was about a week off would be held outside at the Fire House and all those in attendance that day were invited. The Chief's daughter came over and said, "I already have a date for that party but you can tell him that he is my present boyfriend but you were my Ladder". I kind of think that line was also rehearsed because her dad was standing right there grinning. It really did not make much sense because ladder (even though spelled different) sounded like it should be after and not before.

PARTING GIFTS

When I was at my briefing in Washington I had approached one of the Lieutenant Colonels who was also an escort. He was taking a group to Switzerland (which had been one of my choices). I noticed that he was coming from Maxwell Air Force Base in Montgomery, Alabama and I wanted to tell him I was from Birmingham. When I told him, he said he loved being stationed in Alabama. He had been at Maxwell before when he went to the Air War College. He also told me this was his second time to be an escort for the International Air Cadet Exchange Program. He said that he had wanted to meet me when he saw that General Monteith had been the one who selected me. He said that he used to work for the General and added "You two Lieutenants must have really impressed him because he's a tough old bird and not pleased easily". I decided not to tell the toilet paper story.

I did ask him for some tips about being an escort since he had done it before. He said, "Where you are going I'm not sure what you might run into. You know they don't get along too well with their neighbors". He also added "I see you are being given the most petty cash of any of the escorts, so they must think you are going to need it". All the Air Force Officers (instead of the Civil Air Patrol escorts) were in charge of expenses. We were given cash. I saw what the Colonel meant when I received a large envelope with several thousand dollars. He did give some good advise when he said you need to take some gifts to give your hosts. You will have several people at your destination who will be in charge of taking care of you and your cadets. I asked what he was taking. He said that there was a souvenir shop around the corner from our hotel. He said they have some nice things. He had bought paperweights made like a scale model of the White House. I asked how many gifts I should take. He said probably six would be plenty.

When we had a break, I walked around the corner to the shop. I saw the paperweights. They had some made like the White House, the Capital, the Washington Monument and the Lincoln Memorial. They were all so heavy I thought six would be too many to try and pack. The Kennedy Silver Dollar had just come out. They had some of those mounted in silver circles with a key ring attached plus they were in gift boxes. I had heard that President Kennedy had been popular in Israel so I bought six.

Now we are getting close to the end of our trip and I start counting up how many hosts we have had, plus the people like the Colonel who invited me to his son's bar mitz-vah. I had given his son one of the Kennedy key chains so I would need something else for him. The Colonel had said his son liked my gift better than any he received so I knew the other five would be appreciated. Since the Israeli Air Force Officers had been tag teaming taking Colonel Mitchel and I out at night my list was long. I called the U.S. Air Attache at the American Embassy and asked if he had any suggestions. He came up with a good one. He said, "You do know that you have access to our package store here at the Embassy. People in Tel Aviv have trouble finding Scotch. Why don't you buy a few fifths. Since they are hard to come by they should make great gifts".

I talked to Colonel Mitchell. No one had warned him about bringing gifts so we decided we could give the bottles of Scotch from both of us and split the cost. I mentioned that not everyone liked Scotch (I hated the taste) but he said, "If it is so rare around here, I bet everyone would like to have a bottle to serve to guests". So after we started counting, I went to the Embassy and bought a case and a half of Scotch.

I wrote a little about the amount of cash I was given. I had worried about carrying such a large sum and so far everything the Colonel and I did at the hotel, meals, drinks , etc. we had charged to our room. I had asked the hotel manager to put the cash, less my own money and a couple hundred bucks to lock it in his safe, The Passports for the Colonel and I were also there. We had been told we would not need them. Other than showing it at the Embassy when I first checked in I had never even looked at mine. I got it out of the safe to go buy the booze but since the Air Attache met me at the door no one looked at it there either.

UNUSUAL GUESTS AT THE HOTEL

When you stay at a hotel as long as we did you begin to be noticed and you begin to notice others who are also there for a long period. Even though we spent three nights in a hotel in Haifa we never gave up our villa at the Ramat Aviv. It was a good thing too because the Colonel and I had both accumulated souvenirs to take back home so packing them all up would have been a chore.

One guy we saw a lot was the man I played tennis with who worked for Paramount. I played tennis with him twice. We saw him every morning when the Colonel and I waited for our tour bus. He would usually be picked up before we were by a limousine. We never saw him eating dinner or breakfast and he was never in the bar at night, There were two girls in one of the pool villas near ours. They were both attractive but one was much prettier than the other. They were often at the pool wearing bikinis. One day when all of our cadets came over for a pool party. We built what we called an International Pyramid since we had guys from six countries. We made it all the way to four high but when we tried for five it all collapsed. Those two girls had watched and I think our cadets watching them is what caused the collapse of the pyramid.

The next day the two girls approached me. They had seen me with the guy from Paramount and wondered if maybe I worked there too. They wanted to be discovered. They knew about the movie being filmed in Israel and asked if I could get them introduced so maybe they could work as extras. I told them that I had only played tennis with the guy and that I was with the bunch they saw me with the day before. They asked about my Polaroid camera. They had watched as I took photos of our group and had developed them on the spot. Polaroid cameras had not become popular yet in Israel and were still somewhat of a mystery. They asked if I would take some shots of them in bikinis for them to have …… saying they would be happy to pay for the film. Now being the Officer and Gentleman that I was, how could I turn down a request like that.

I used up two boxes of film. They changed bikinis a few times. We attracted the attention of the Hotel Manager and he sat and watched as they posed around the pool. I kept a couple of the photos but gave them the rest (no charge for a pleasant assignment). The really good looking one gave me a small head shot photo of herself and I wrote her name on the back of it. Later the Hotel Manager and I talked. I wanted to know what exactly was the reason they were there so long. He told me that one of the girl's dad was wealthy and had paid for them a two week stay at the hotel. He then added "Well he was old enough to be her dad". And he winked.

THE SEQUEL TO THE STORY

Close to two years later, three Lieutenants and I had rented a house. We were lounging around and one of the guys was reading (or should I say looking at pictures) in a copy of *Playboy.* All of a sudden he said, "Hey Capps there is a whole section in here called "THE GIRLS OF ISRAEL". See if you know any of them. He thought he was being funny but when I looked there she was … the good looking one. She was "more or less" in the same pose as one of my Polaroid shots. It was "More" of her and "Less" of the bikini. The guys all took a look. I ran to find the two Polaroids I had kept. Everyone looked at them and one said "The face looks similar but this girl has lots bigger boobs. it's not the same girl". Then I showed the small head shot photo she had given me. When they turned it over and saw the name on the back, it was exactly the one in *Playboy*. I said maybe her "Dad" paid to have a little something added. True story … as I said before ……. You can't make this stuff up.

Two more guests which we had seen a lot of was a mother and daughter. I had noticed them at the reception for General Weitzman but had not gotten to meet them. The mom and the General seemed to know each other and I saw them together several times that night. The daughter was not a beauty but attractive and always dressed up when we saw her …. usually it was she and her mom at breakfast. One day I talked to her by the pool. She said, "I don't like to swim when the pool is crowded". I told her that I normally would swim early in the morning before I would leave for a tour. I had explained what the Colonel and I were there for and about my responsibility for 15 cadets. Before she left she said that she might join me in the morning at the pool. We met many mornings at 6:30 a.m. We were the only people at the pool except for the guy cleaning it.

I was curious about who she and her mom were and their connection to the General. One day she and her mom were leaving breakfast about the time we were getting ready to board our bus. My Colonel friend was going with us that day and he went over to speak to the mom and daughter. I sat with him on the bus and told him I had been meeting the daughter in the mornings and that other than telling me her name she didn't talk about why they were there. He said "Quite a story. Her mom owns several businesses in Tel Aviv and she may own the buildings they are in as well. She and her daughter live in Berlin. The daughter speaks Hebrew, German, English and some French. She is usually with her mom in case she needs an interpreter. To say they are wealthy would be an understatement. The husband died three years ago. He was an arms dealer. Most of the weapons he sold were made in France. The Mystere that Colonel Mitchell got to fly in came from France and I'm sure her husband negotiated the sale. He sold many arms to Israel and I think the mom may still be running that business as well". Now it was beginning to make sense why she had been talking to the General.

The daughter and her mom were staying in a suite on the second floor of the hotel and their balcony overlooked the pool. That's how she knew when I got to the pool on the mornings that I had time. One morning she came down. A big group of kids had arrived the night before from Ireland. They were loud when they arrived and now at 6:30 in the morning they had all come to the pool. The daughter said let's come back tonight late. I told her we were visiting a Kibbutz that afternoon and I did not know when we were supposed to be back. She said that she would call my room. Speaking of calling me she had been pronouncing my name Beal (with emphasis on the long "E") instead of Bill. One of our cadets heard her one morning when she said goodbye as we were getting on our bus. From then on the cadets were calling me "Lefttenant Beal" and really stretching out the "BEEEEAL". They did it the way her accent had pronounced it and would all laugh.

The Kibbutz was interesting. It was kind of a collective community where they had a huge farm. I think at the time there were quite a few of these in Israel. We had lunch there and they gave us a tour. A woman that I met there asked if I was staying with a family like all the cadets. When I said no, that the Colonel and I were staying at the Rmat Aviv Hotel, she said that she wanted to show me a photo. She went and got a photo of a group and pointed out one particular girl. She said, "She stayed with us all summer. She is an American from New York City. She is staying with a family for her last week in Israel and they live close to your hotel. You should call her up. Maybe the two of you could go out. One of our hosts, a Major who's family I had gotten to know well overheard the conversation. When she went to get the phone number of the family the Major said "She sounds like a "Shadchan" to me". He said"Your grandfather would probably say "Yenta" as it is more Yiddish". I had told his family that my mom and grandfather were both Jewish. I said that I had not heard of either one of those words before. He just smiled and said "Matchmaker".

The daughter and I went to the pool around 11:30 p.m. and stayed until 1:00 a.m. This became a nightly ritual for about a week. One night she said "Beal we are compatible. I have much money and I want to travel around the world while I am young, why don't you and I go see the world". I thought it was a joke and I said "Sure that sounds like fun". I had found out that she was 33 so to me at 24 she didn't sound that young but I thought no more about her comment until one morning at breakfast. Her mom had gone back to Berlin for a couple of days and her daughter joined the Colonel and I. In the middle of breakfast she said to Colonel Mitchell "Did Beal tell you he and I are going to travel around the world'? She actually sounded serious. I almost choked and I think I spit orange juice because I had just taken a sip. As Lucile Ball or Ricky would have said, I had some "splaining" to do. I said that I was committed to several more years in the Air Force and although I would love to see the World it would have to wait. The Colonel could not wait to tell that story to our hosts. Did not see much of the daughter after that.

I called the family where the girl from New York was staying. The group photo was not that clear but from what I could see she looked like she had a great smile. I told her who I was , why I was in

Israel and who suggested I call her. I invited her to dinner at the hotel and she accepted. We danced in the bar after dinner. They had a band and we had a good time. Our farewell party was coming up at the end of that week and I really wanted to have a date for it. I would have invited the Air Attache's daughter since she was my date for our first party but she had gone out of town and would not be back before my last night.

I had considered asking the good looking bikini girl but she seemed too interested in herself. Now it appeared the Arms Dealer daughter was out because I could not go around the World with her. The Fire Chief's daughter had already told me she had a date and besides she was too young anyway. I had thought quite a bit about giving a farewell speech. I wanted at least to give part of it in Hebrew. I really had not practiced what little bit of it I had learned because everybody had been speaking English. I called the girl from New York. We went out three more times that week. When I told her what I wanted to do at the party (I had already asked her to be my date) , she said " You write your speech and I will translate it into Hebrew. If I have a problem with any of it the family I am staying with can help. We will write it in Hebrew and I'll help you memorize it".

She was beautiful and the night of the party at the Fire Brigade she looked terrific. We had practiced and I stumbled thru my speech okay but she said "Look I will be right next to you … just speak slow and if it looks like you are having a problem. I can coach you. I do it all the time when we are having a play in school". She was a student at NYU.

SHALOM AND L'CHAIM

Phonetically in the U.S. you usually hear L'Hayim but it is a toast to life. Hebrew letters are also used as numbers and the Hebrew letters that spell "chai" stand for 18. That is why it is considered a lucky number in Jewish culture. My farewell speech gave 18 reasons why I loved my visit. Many were the names of all those who acted as our hosts. I am not sure how well my Hebrew sounded but they got the gist of what I was trying to say and at least they were polite enough to give a standing ovation ….which was easy because we were all already standing in back of the Fire House. My date said, " You spoke Hebrew with a Southern drawl ….not bad". I liked her a lot and would loved to have seen her again. We wrote for awhile but by the time I got re-stationed to Pennsylvania and made a trip to New York we had lost contact. Where was that matchmaker when I need her.

HEADING HOME …. WITH EXCITEMENT

Going back I of course was in uniform. I told Colonel Mitchell I would go and settle up with the hotel. I figured we had run up quite a bill because we had, in addition to eating many breakfasts, dinners and some lunches and often with a guest we had also gone to the bar more than a few times. Then there were dry cleaning bills. I had not worn my uniforms too many times but it was so hot there they had to be cleaned each time. All that with adding our room cost should knock a dent in those thousands I had stashed in the vault. When I got on the bus the Colonel said, What was the bad news"? When I said zero ….our total bill had been paid by the Israeli Government. Since we had no emergencies, I was going to be able to turn in the majority of what I had been given. I gave the Colonel back his half of the case and a half of Scotch. We decided it could fall under miscellaneous and it was not that much anyway.

Flying back the Israeli Air Force took us to Greece instead of Turkey. We met the Athens delegation and the same two German pilots with the General's plane. We flew from there to Turkey. When we landed we were in for a shock. On the taxiway we were met with a truck with flashing lights and they ordered the plane halted. I was sitting in the cockpit and suddenly saw a truck pull up next to the plane. Soldiers all jumped out with weapons pointing at the plane as they surrounded it. A small tractor came up pulling a set of stairs. A soldier (I think a Turkish Captain) came to the door with a pistol drawn and ordered (I guess in German or French because the pilots and my two French cadets understood. I did not)…..ordered us Off The Plane With Hands On Our Heads. We did as we were told. Other than the German pilots, the Lieutenant who had been the escort for the Athens cadets and I, were the only other ones in uniform and they wanted us to step forward. I was not sure what was going on but it did not look good. I told the cadets to not do anything because those machine guns are probably loaded and we don't want to give anyone an excuse to shoot. The other Lieutenant said the same to his group. Both CAP officers who had been sitting together were in agreement.

Two of the soldiers started unloading all the luggage and lining the suitcases along the taxiway. All of a sudden I see a jeep coming up fast. There is an Air Force Captain and what looked like an Air Policeman driving. The Captain hopped out and said "Lieutenant do you have a set of orders"? I had kept a set in the pocket of both my dress blues and my khaki's, which I had on. I also handed him my Passport. That was when I first learned I had been traveling with a Diplomatic Passport. It read "Abroad On Official Assignment For The United States Government". He said, waving the Passport "This is Gold". He went over and talked a long time to the Turkish Officer. Apparently the Air Policeman was also an interpreter. They started loading our bags back on the plane Our cadets and the two CAP Officers did as well. The Captain said "I think I can explain the mix up. Your counterparts, the CAP cadets who visited Turkey are waiting in Base Ops. While you were in the air a disagreement occurred over Cyprus between the Turks and Greeks. Fearing a conflict they went on alert. When someone saw a military plane coming from Athens to here they wanted to take no chances. They knew it was a German Air Force plane but decided that just to be safe they would stop and search it. Everything is cool now .... taxi on to Base Ops and pick up your passengers. I had a story to tell Lieutenant Ahart, who had been with the group in Turkey.

Nothing came of whatever conflict they avoided on Cyprus that day but 10 years later (1974), Turkey did invade the island and met resistance from Greek and Greek Cypriot forces.

"FIGMO"

I won't explain what the first two letters stand for but you can figure it out when I tell you the last three mean "GOT MY ORDERS". Before I left Amarillo for my long TDY trip to Israel, my paperwork had gone all the way through channels requesting that I be reassigned to Recruiting Service. Also before I left, Colonel Parker called me to his office and said "Congratulations you pulled off another one. You got orders to report to a USAF Recruiting Group in Pennsylvania. You are going to Olmsted Air Force Base in Harrisburg, Pa.". The orders were dated 24 June, 1964. I did not know it at the time but that date would become significant because I would eventually be married on June 24 in Harrisburg, Pa. I only discovered this while researching old orders when writing this book. An "Omen" …. it may well have been.

Normally when anyone in the military say they are FIGMO it means they sort of slough off and take it easy for the rest of their current assignment because they know they are about to leave. I knew so early and besides I had my temporary duty trip to Israel in between. I was not to report to Recruiting Service until the first of September. Before I left on the TDY trip I had turned over the Base Beautification duties entirely to my NCO from the Retraining Group. By the time I left he was doing everything anyway. He still drove the Army National Guard weapons carrier and no one said a word about returning it. I had him sign a "Hand Receipt" in place of mine. Besides he had carried a few more cases of toilet paper to the Sergeant at the Guard Headquarters. I never figured why their supply channels did not work.

The squadron had been running smoothly while I had been gone. Major Gohagan and Sergeant Summers hardly missed me. It brought to mind what one of those old Senior Master Sergeants on my Quasi-Board Of Directors once told me. He said, “Lieutenant if you ever think you are indispensable just stick your arm in a bucket of water and pull it out. The hole you leave is what you really are to keeping the Air Force going”. He was the one who always had a knack for telling it like it is.

DETAILS

A friend of mine asked how I remember so many details from so long ago when I was in the Air Force. I kept good notes, especially when I went to Israel. I knew back then, that particular trip was special. The State Of Israel was only 16 years old at the time and it had always been in the news. With my Jewish heritage, I knew there was a reason I was chosen to go there. Also I kept a copy of every order I ever received in the Air Force. I think I knew that someday I would write a book. After all, my High School English teacher (Miss Dannelly) had predicted it. Looking back over all those orders has refreshed my memory of the significance of each one. I have kept them in a filing cabinet in my basement. We had a flood and some of them got wet but they are still readable. Getting them in order by date was no easy task but it has made the flow of this book easier to understand.

MEANWHILE BACK IN AMARILLO

The last couple of weeks went really fast. I had a lot of packing because over 18 months I had accumulated much more than when I first arrived. I needed more than just my convertible to move my stuff. I am sure the Air Force would have paid to have it shipped but I did not want to do that. I saw an ad for a used one wheel trailer. I had never seen one of those. It was a fair sized bed with one wheel under it, in the center. Two clamps fit over the rear bumper of the car (remember cars back then had bumpers). I bought it. The guy from whom I purchased it clamped it on my car. Empty, it rolled pretty good going back to the base. Before long, I had it really packed. I had put my most expensive items, the black and white portable TV and my stereo in my trunk. I had built a bar which had a pull down door which doubled as a drawing table to do my monthly calendars for the Officer's and NCO Clubs. It was large and went in the trailer as did pots and pans and my toaster oven, a couple of lamps and a night side table and two boxes of art supplies. It was loaded ….probably too much. I covered it with a tarp and strapped it down. The back seat and front shotgun seat had boxes of my clothes and other junk. I looked like the Beverly Hillbillies heading to California except I was going the opposite direction.

Funny thing about leaving a place where you have been so long. Things happen just about the time you are leaving that you wish had happened sooner. Mike Rex and I had attended a basketball game at Amarillo College. There was a Head Cheerleader who caught my eye. I started asking what her name was. I found out she had been Miss Amarillo. Somehow I got an introduction and asked her for a date. She said she was going steady but said give me your phone number and if we ever break up I will give you a call. Sounded like a cold shoulder to me but I gave her the number of the pay phone in the BOQ. She would not give me her number. This had been over eight months ago. At around eight p.m. on the very night I was signing out at midnight the phone rang. Since I was running back and forth packing my car I happened to answer it. Yep, it was her. She said, "My boyfriend and I just broke up . Do you still want a date"? Just wasn't meant to be.

There was this thing about going on leave or traveling to a new assignment. If you signed out at one minute after midnight you saved a day. Most Junior Offices always did this …. We were young and could drive all night. My orders gave me seven days travel time to go from Amarillo, Tx. to Harrisburg, Pa. I had figured driving time to be a bit over 24 hours. My logic back then was that I had plenty of time to drive to Pennsylvania by way of Birmingham and see my parents. I pulled out around 12:30 a.m. Driving by myself all night I was wearing a gun. Lieutenant Colonel Angus had sold me the Belgium Browning 9mm Automatic pistol which he had been issued when he flew with the RAF. With a teenage daughter his wife did not want weapons in the house so he wanted me to have it and a British .303 rifle. I still have the automatic. It has become a collector item.

Not 40 miles outside Amarillo I had a flat on my car. Getting to my spare and jack was not easy. I almost had to unpack the trunk. The flat was on the front right tire so I figured I could change the tire without removing my one wheel trailer. I was having a difficult time until an eighteen wheeler stopped and the driver helped me. He said he thought I had probably come from the base. After we changed the tire I followed him. When he stopped at a truck stop I did also and bought him a cup of coffee and a piece of pie. He was a prior service Marine. He suggested that I stop every so often and make sure the bumper clamps were tight because they were the only thing keeping my trailer upright.

I made it to Birmingham and had only stopped one night at a motel. I stayed three days and still had plenty of time to drive to Harrisburg. I pulled into a small town called Highspire, just outside of Harrisburg. It was next to Middletown where Olmsted Air Force Base was actually located. It was close to six at night and it was pouring down rain. I was looking for a motel because I did not want to try and sign into the base at night. I turned a corner and my one wheel trailer came loose from one side of the bumper and flipped over. It was a mess. The bar I had made was broken and other stuff was scattered all over the street plus I was getting sopping wet. I must have looked like a lost ball in high weeds and a wet one at that. A man pulled over, hopped out wearing a raincoat and said "You look like you can use some help". Did I ever. He was so nice. He had tools in his car and between the two of us we got the trailer upright and fastened back to the bumper. The clamp on the other side which had not come loose was now bent. He straightened it as much as he could with a hammer. Together we piled my stuff back on the trailer. The tarp was torn.

He was Polish and his name was really Polish sounding. He said you are coming home with me. You can change into dry clothes and I will show you a good motel. I had told him who I was and where I was headed the next morning. When we got to his house which was just around the corner from my accident he introduced me to his wife and 22 year old daughter who just happened to work at the base. I went to the bathroom and changed out of my drenched clothes. When I came out , dinner was being served and his wife insisted I stay.

After dinner, his daughter had me follow her to a motel which was just outside the base. As I was checking in I met another Lieutenant (Jack Lausten) who was doing the same as I, waiting to check in at the base the next morning. We stayed friends the whole time I was stationed at Olmsted. The next morning when I signed in I asked if I had to live in the BOQ or could I live off base (which I really wanted to do). They looked at my records and saw that I would be promoted to First Lieutenant on the 26th of September. They said with your date of rank you can have a choice. You can live on base in the BOQ or get a housing allowance and live off base, I knew I would be getting a raise as a First Lieutenant and if I got a housing allowance it would be even more if I chose to live off base.

Now I had an immediate problem. Where could I find a place. Before I checked out of the motel I had said that I might be back for another night since I had no idea what to expect when I got to the base. I had signed in early so I had at least two days before I needed to report to the 3502nd Recruiting Group. Driving around downtown was not easy since I was pulling a trailer. I found a place in Middletown, which was just outside the base to park and leave my car and trailer. I thought

I would walk around downtown and explore. First place I went was to a bank to open a savings and checking account. I asked if anyone there happened to know of an apartment I could rent. One older lady said she had a friend a few blocks away who did not have an apartment but she thought she had a room she wanted to rent. I got the name and address and walked there.

When I knocked on the door a little old lady answered. I told her what the woman at the bank had said and she replied that she indeed had a room she needed to rent in the back of her house. We walked down the alley at the side of her home. We went in a back door. She said that her husband had died two years before. She had been a little strapped for cash. Her son, who was married and lived in Harrisburg suggested she rent out a room. He had taken the back part of her house where there was a half bath and closed off two rooms. In what used to be a closet he added a shower. In the room next to the half bath which now had a converted closet were two single beds. With no closet there was a rack like you see in a store to hang clothes. The door going toward the front of the house just had a sheet of plywood nailed over the opening. She said, “My son is not quite finished yet. We haven’t even put up a sign. I had noticed a hand written piece of cardboard when we came in which said ‘ROOM FOR RENT” in the far back room where the back door was. He was in the process of converting it into a small kitchen. It already had an old refrigerator and a small two eye electric stove. There was a sink on the floor but not installed. I'm looking….. and instead of a room for rent I see an efficiency apartment. It sure had more room than I had in Amarillo. She said, "I don't think it will be long before he finishes so if you are interested the rent will be $60 a month ….paid in advance. My son said I needed to collect in advance in case someone skips out". I asked about parking. She said, "Well I don't have a car so you can park in the back". There was no back yard but room for my car. I could lean my trailer after I unpacked it up against a fence separating the house which was facing the other way at the end of the alley. I said that I would like to take it (as is) today if I could. She said , "Great I will call my son …. he will be happy I found an Air Force Officer to rent it to". She knew because I was still in uniform. I handed her $60 cash and started unpacking. Timing is everything.

I did without the kitchen sink for my first month but it was not a problem. I had water in the sink in the small bath. I met her son that night. I guess he wanted to make sure his mom had not made a mistake renting to the first person who knocked on her door. He seemed to approve especially when I said I would help him finish off everything.

FIRST NIGHT AT THE OFFICER'S CLUB

I felt pretty good. I had already secured a place to stay and unpacked. I was not settled in but decided to celebrate by going to the Officer's Club for dinner.. The first person I run into is Jack Lausten who I had met the night before. He had moved into the BOQ. I told him about my place and he said he did not have enough time in grade to live off base. Olmsted was his first assignment. We met several Junior Officers. I spotted the daughter of my Polish Rescue Squad sitting at the end of the bar surrounded by a bunch of guys. One of the Lieutenants Jack and I had been talking to said, " I need to introduce you two to the girl at the bar, she has a reputation for sleeping around". When he started to introduce us she said, "Oh I already know Bill, we had dinner together last night before I took him to his motel". To say I received a few looks and was immediately tapped as a fast player would be putting it mildly. During my stay at Olmsted she and I became good friends and I introduced her to one of my eventual roommates and no, we never dated.

THE 3502 RECRUITING GROUP

The next morning I reported into the recruiting Group Headquarters building. The fist person I meet is a Captain L'Italien the Chief Of Personnel and Administration. He says "Welcome to

Recruiting I understand you are the first Second Lieutenant to be assigned. I also noticed you were an Administrative Officer. That's good because you will be my backup when I'm gone". I was not exactly sure what that meant but I was pleased when he next said I would be the Information Officer for the Advertising & Publicity Division. That was what I had wanted. The next person he introduced me to was Colonel Thomas S. Torresson, Jr. the Commanding Officer. He seemed extremely nice, he welcomed me and took me to the building next door (which was smaller) to meet my boss, Captain Paul Buchsbaum who was Chief of Advertising & Publicity for the Group. I did not have an office but was given a desk which had partitions around it. He introduced me to the Advertising staff. We had three NCO's and an Airman First Class who was an artist and the Captain also had a civilian secretary.

I had a lot to learn about how a Recruiting Group operated. We were set up to function like a small base. We were independent of most things at Olmsted. It supposedly was that way in case we had to keep operating off base. We had our own Personnel Department, our own Motor Pool and our own Operations Division which controlled two airplanes which were used by the Group. Several different divisions had pilots who alternated flying our planes. The planes were called U3A's and were like twin Cessna's. The Motor Pool had quite a few

Government Vehicles which were spread over seven states. This also included a panel van which was mostly used by the Advertising staff. At Olmsted, the 3502nd Recruiting Group was considered a tenant unit.

The 3502nd controlled recruiting in seven states which included Pennsylvania, New Jersey, Delaware, Virginia, West Virginia, Maryland and Kentucky. There were only six other Recruiting Groups in the U.S. In each of our states we had a Detachment Headquarters, each with a large staff. The Detachment Commanders all reported to Colonel Torresson. Recruiting offices all over each state reported to the Detachment Commanders. It was a big operation.

OLMSTED AIR FORCE BASE

Olmsted had been around a long time. It was first built in 1918 and was known as Middletown Airfield and used by the U.S. Army Signal Corps. It was used as a supply depot for years and when I got there it still was being used that way. The name changed to Olmsted in 1948 when it was named after a 1st Lieutenant who died in a hot air balloon race. The base was not as big as Amarillo although the size was deceiving because it was split, with part on one side of the road connecting Middletown to Highspire to Harrisburg (which is the State Capitol of Pennsylvania). The higher side had many warehouses where the supply depots operated. We were on the side which was parallel to the Susquehanna River. Our side had the Officer's Club, the Flight Line and most of the main operations of the base. There were beautiful homes for Senior Officers along the runway which was also parallel to the river. Colonel Torresson and his family lived at the one on the end right next to the Officer's Club. I found out early on that he was a Command Pilot and had more time in grade than any Colonel on the base. Olmsted would eventually close in 1968 or '69 and is now home to the Pennsylvania Air National Guard and the Harrisburg International airport.

EARLY TRAVEL

I found out fast that my job would involve lots of travel. Being single and wanting to stay busy this suited me fine. Before the end of the second week the Captain and I were flying to Newark, New Jersey and New York City for meetings. It would be the first of many trips I would make in one of the U3A planes. I had only flown once in a small plane (when I went up in the glider in Israel). It was scary at first but I soon got used to it. Later on I was involved with two near misses but I will touch on them later.

While in New York I tried to call the girl I had met in Israel. We had not written each other in awhile and I never made contact. One of the meetings in New York was with the Chief Of Advertising and Publicity for the Group covering that state. Since none of us were competing in recruiting (since it was by state), we often swapped ideas with the other six Groups. Because we had a Detachment Headquarters in Newark, New Jersey we did compete for Advertising and Publicity contacts in that metro area. Of course any publicity benefited both Groups. The word CONTACT became important and the Captain stressed that to me early on. We actually called them COI's (centers of influence). We pushed all recruiters to keep a list of them. Back in the sixties none of the military branches had a budget for paid advertising. Our main job in our Division was to try and get as much Public Service advertising in Radio, TV, Newspapers and Magazines as we could. We competed with the Army, Navy and the Marines constantly.

By the 26th of September I was promoted to First Lieutenant. I had already made three trips, the other two by car. Our military vehicles were nice new Air Force Blue sedans with the large round Recruiting Seal on each front door. The Recruiting Seal was also worn on our uniforms. It was a two and one have inch ceramic gold rimmed circle with the Air Force emblem in the center. They looked good on the uniform. I did not have many ribbons so I liked wearing the Recruiting pin.

GETTING ACQUAINTED AROUND BASE

By the middle of October, 1964 we were starting to get more Lieutenants assigned to Recruiting. They were all First Lieutenants. Two of them were from Amarillo. They had copied my application and just changed the name. One was my old buddy, Jake Laeufer. When he got there they gave him the option the same as they did me ….live in the BOQ or live off base. I talked him into living off base. He did not have a place and I had two beds, so he moved in with me. We thought we could manage the small space until we could find something bigger.

Even though Recruiting was a Tenant Unit, I still got assigned a few Extra Duties on base. One was extremely enjoyable and allowed me to meet many new friends from other parts of the base. I was appointed to be a member of the Junior Officer Council. The first two guys I met worked right next door to our main building. Their names were Dell Fisher and Charles Earnhart. They were both OSI Agents (Office Of Special Investigation). They wore civilian clothes and their rank was supposed to be secret. Of course being the age they were, everyone guessed they were First Lieutenants. Thru the Junior Officer's Council and the Officer's Club I soon had contacts all over the base. One happened to be the First Lieutenant who was Manager Of the Officer's Club. Even though I was traveling a good bit of my time I could not resist trying to make a little extra money. I showed him samples of the calendars which I did for the "O" Club in Amarillo and I soon started doing them at Olmsted. Using their Mimeograph Machine to print those calendars would ironically become one of the best decisions of my life.

ORIENTATION GROUP

Recruiters out in the many Recruiting Offices we had throughout our seven state area were always in the need of an Air Force display. They would often need stuff to decorate a booth at a school fair or various functions. If they knew far enough in advance (which most of the time they did not) they could have our office order something from the Air Force Orientation Group. They were stationed at Wright-Patterson Air Force Base in Ohio, where the Air Force Museum is located. They had plenty of items which could be exhibited but it took time to get them sent all over the country. One of my first trips with Captain Buchsbaum was to Wright-Patterson. We had a meeting there and I met many people who were from the other six Recruiting Groups. When we came away I remarked that The Recruiting Groups closest to the Orientation Group had an advantage over us because they could get displays quicker. I asked why couldn't we just get some of our own. The Captain thought that sounded like a good idea but he said he did not know how we could do that. With the "Dog Robber" mentality I had developed in Amarillo I was already thinking outside the box. It's not always what you know … it's who. Before we left Ohio I had met a civilian (I think he was GS 12) who was pretty high up in the chain of command with the Orientation Group. I got his phone number and asked him if I could call him if I had questions about displays.

When we got back to Olmsted, I talked to my new contact several times. I found out that the museum had so much stuff that they did not have room to display that it was just stored in a warehouse. My first acquisition I think shocked everybody. When I started asking what they had that I could display, even in a Recruiting Office, he said that he thought a Sidewinder Missile might look pretty neat. It was originally used by the Navy but the Air Force had also started using them in 1964. I found a photo of one and it looked sharp. He said that there were bunches of those that had originally been planed for the museum. The rocket and explosives had never been installed. I asked if I could get some on a hand receipt. I figured if I could get a truck that way why not some missiles.

One day a C-130 cargo plane arrived on the flight line. They called and said we have a shipment for a Lieutenant Capps. I now had 20 Sidewinder

Missiles. We divided them up between our seven Detachments to distribute as needed to their Recruiting Offices. That was the start of what eventually became known as the Capps Orientation Group.

By April of 1965, Captain Buchsbaum had been reassigned and I took over as Chief Of Advertising And Publicity. I originally thought it was going to be temporary until they could get a replacement but Colonel Torresson told our new Chief Of Personnel to make my new title permanent. My displays started to increase. When I would go out in the field to visit Recruiting Offices they kept asking if we could get something big for them to have when they needed a display for a special event.

My staff and I kicked it around. The Space Program was getting a lot of press. Why couldn't we build a model of a Space Capsule, put it on a trailer to pull behind a Recruiter's car (most all of them had a trailer hitch). I had made some contacts in the Aircraft Maintenance shop. My staff artist started drawing up plans, My contact in Ohio sent me a control panel. It came out of a plane but we were just going for looks not accuracy. He also sent me an ejection seat from a jet. Somebody donated a male mannequin. We got it dressed in a flight suit with a helmet. We built (with a lot of help) what looked enough like a Space Capsule to attract attention. I got so many different people involved (all obviously not authorized) that our creation could be described as "Built by Committee". Somebody put in a Black Light and painted all the instruments with Day-Glo Paint. With the mannequin laying back on that seat it looked pretty real. When we sent out photos to our Detachments …. requests for its use came flying in. The Colonel was not only impressed with our creation he loved the publicity we were getting with it …..that is until the day it crashed. It had been used at County Fairs, displayed at Armed Forces and Veteran's events, high schools, colleges and many venues. But one day a Recruiter from Philadelphia needed it for a display by the Liberty Bell. He came to Harrisburg to pick it up. He was running late. Going too fast on the Pennsylvania Turnpike he jack knifed the trailer. I got called to the Colonel's Office. He held up the paper. The headline (with a color photo) read "SPACE CAPSULE CRASHES ON PENNSYLVANIA TURNPIKE" It did go on to explain that it was only a display and no one was hurt. I guess I didn't have great luck with newspapers. Aka "Cappsule".

My buddy in Ohio had heard about our Space Capsule as we had sent him photos along the way. He got a big kick out hearing about its earth landing. I told him I needed something which would top it as far as a big display. I was getting a reputation and had caught the attention of the

Information Office at the Pentagon. As the Chief Of Advertising And Publicity my AFSC (Air Force Specialty Code) was now that of a fully qualified Staff Information Officer. There were not that many. I had been to the Pentagon a couple of times and had met the officers in the Information Office. We had a Detachment Headquarters in Arlington, Va. Since we were the closest Recruiting Group to Washington the guys in the Pentagon would call me when they got a request from a Congressman or Senator for an Air Force display. My Ohio contact said, "How big are we talking about"? Remember I had been stationed in Amarillo. We had a Strategic Air Command Wing there. I had watched those "B"52's take off and land. They carried Hound Dog Missiles. I said "How about one of those". He was amused …. but he didn't say no. After a pause he said "I actually might be able to get one". By now the Colonel was used to me being a bit off the wall but I think this one really took him by surprise. I had a date on a Saturday night and did not get back until very late …. keep in mind no communication like today. The Air Police had left a note….call Colonel Torresson. I hated to wake him at 2 a.m. on Sunday morning but the note said to call. He said "I am sure you have a logical explanation but down on the flight line the Air Police have armed guards standing guard on a missile, and I mean a big missile which the bill of laden says deliver to Lieutenant Capps". It had come in on a cargo plane late Saturday night. Hard to believe… but true. I received a Hound Dog Missile signed to me on a hand Receipt. I think I still remember what the receipt was called ...a DD95. You never know unless you ask.

THE HOUND DOG MISSILE

This was either my greatest coup or my biggest mistake. It got a lot of super publicity but also caused many headaches. First of all we estimated that it weighed over 10,000 pounds because even though it was non operational it still had the jet engine. It would not fit on a normal flatbed truck … it was over length. It took the biggest forklift on base to lift it but I now I had to figure how to use it. First I called the Base Motor Pool. Unlike the one at Amarillo the civilian who ran it at Olmsted wanted to help me. He said that he could assign a cab and a trailer but I had to have an NCO or Airman with a valid Government Drivers license, qualified to drive it. One of my NCO's volunteered to go take lessons and get qualified. Next came the loading and unloading. We would only be able to use the cab and trailer when we had an event. Two people needed to be licensed to drive the big forklift. That is how I came to be one of the few First Lieutenants with a Government License to drive a forklift which could lift a 10,000 lb Missile.

Sitting on the trailer, it was spectacular. It was mounted on a base with wheels which could hydraulically lift it up and even tilt it skyward. When it was on the road "Long Load" signs were on the back. Going thru cities we normally called and got a Police escort. This only added to the attention it received. At Fairs we always cranked it up so that it was pointing to the sky. Almost every time it was on display it made the paper. Occasionally not for the best reason. One Saint Patrick's Day it was on display at a State College in Pennsylvania and it got spray painted green. It was not easy but my contacts in aircraft maintenance managed to get it repainted as good as new.

My NCO who drove it always complained about how hard his job was driving it. He kept saying the truck beat him to death. He finally talked me into riding with him when it was being displayed at an event in Upper Marlboro, Md. It was only about a four hour trip and he was right I felt like I had played running back for the Baltimore Colts … with no pads. I could not feel too sorry for him though. We found out that since he was being given cash to buy gas he would only buy it from service stations which gave Green Stamps. He supplemented his income pretty well.

Google a photo of a Hound Dog Missile and you can get an idea of just how impressive it was. In

Israel I hung out with the daughter of an Arms Dealer …. now I was beginning to feel like one myself.

NEW PERSONNEL OFFICER

When Captain L'Italien was reassigned I filled in as Chief, Of Personnel and Administration. With trying to do both jobs it meant many late hours. That office was in our main building right next to Colonel Torresson's. Because of this I saw him every day. We shared the same civilian secretary so he was aware of how much typing I required her to do. I had an advantage on the Administrative side of the job because I had cut many orders when I was in Amarillo. There were so many more in Recruiting because every Officer, NCO and some of our Airmen in all of our Divisions were constantly going TDY to our Detachments and Recruiting offices in seven different states. On the Personnel side I had to study regulations. I was happy when the new First Lieutenant, Howard Glick showed up to take the job. He and I hit it off from the start. He was going to make the Air Force his career and he was already Regular Air Force and not Reserve like I was. I think the reason we hit it off so well is he had that Dog Robber instinct like I did. Together we got things done and not always by the book. On some things it almost became a game especially later on when we had to change bases.

ADVERTISING & PUBLICITY SERGEANTS

At each of our Detachment Headquarters there was an NCO who handled Advertising and Publicity. They did not report to me but I had direct contact with all of them. They reported to each Detachment Commander. The normal tour of duty in Recruiting was three to four years. Some of the NCO's who were in those positions at the Detachments had been there longer because of their contacts with the media.

Our NCO in Charleston, West Virginia (Sergeant Harmon), I think they said he was going on his eighth year. He was originally from that area and knew everybody. He was fun to be around. He had me fly down to attend the West Virginia Broadcaster's Convention in Charleston. He must have known every disc jockey in the state. He had an Air Force booth set up at the convention and had several of the Detachment Recruiters manning it. He and I were walking around talking to different exhibitors and attendees. The booth right next to his was the Telephone Company. He asked if he could use their phone to let the guys at the Detachment know I had gotten into town, We had come straight to the Convention from the airport. When he got on the phone he said "Let's take Lieutenant Capps out tonight and see if we can get him bred". What he did not know until those in the

Phone Company booth started telling him was their phone was hooked into the speaker system for the whole arena. People were laughing everywhere but he quickly recovered saying loud into the phone "AT THE BAKERY, AT THE BAKERY" Walking around the rest of that day with my name tag clearly visible brought lots of comments …. like "Hey Lieutenant Capps, see if they have cake too". You can imagine what a group of disc jockeys can come up with.

Another NCO who was really good at coming up with publicity ideas worked in our Detachment Headquarters in Newark, New Jersey. He set up a publicity stunt with the New York Mets. Yogi Berra was their Manager at the time. A recruiter from one of the Jersey Recruiting offices had a birthday on the same day as Yogi Berra's. Our Sergeant Gollara set it up for us to go on the field during batting practice for a game in Shea Stadium. We had a big cake so we could do a photo op with Yogi and the Recruiter. We got there early. I was not a big Pro Baseball fan but I knew who Yogi was and was looking forward to meeting him. The first person to greet and welcome us was Ron Swobota who played outfield. He said, "Lieutenant Capps do you want to take batting practice with us"? Now what 25 year old guy is going to say no to a question like that. I took off my uniform jacket and stepped into the batting cage. The first pitch I hit a line drive right over second. The next pitch I took a big swing and hit it all the way to the outfield. Now I'm thinking I can do this...it's not as hard as it looks. I should have been suspicious when Swobota got a bunch of the players to come gather just outside the batting cage.

I never saw the next few pitches and the last one looked like it was going to hit me and curved right back over the plate. I went to the ground on that one. They were all laughing. I had been set up and he was only the batting cage pitcher not even a regular player on the roster. I did at least get my picture taken with Swobota and a couple other players. The highlight of the day, however, was most unexpected. Not only was it Yogi's birthday it was Casey Stengel's last day and he came on the field to meet us. We got to sit in the box seats next to the dugout. The Mets lost that day but the Recruiter managed to recruit the bat boy to join the Air Force.

We had another photo op with Dean Martin's daughter Dena, when she was performing at a nightclub in New Jersey. Sergeant Gollara pulled off several events which garnered lots of publicity. He was a friend of Jackie Gleason and we heard he went to work for Gleason Enterprises when he retired from the Air Force.

The wildest publicity stunt which we pulled off

was one I was involved with in setting up with our Advertising and Publicity NCO at our Arlington, Va. Detachment. They got several of their Recruiters to put together what was called a Buddy Flight where several kids are sworn in together while joining the Air Force. With a lot of help from many different people we were to swear in eighteen, eighteen year old's on the eighteenth Anniversary of the Air Force at the Air Museum in Washington, D.C. in front of a plane which was eighteen years old. They were to be sworn in by the Secretary Of The Air Force Eugene Zuckert.

I had to drive down from Harrisburg to Arlington the morning of the event. I left before daylight and I guess being half asleep when I put all the stuff on my fresh cleaned uniform I forgot to put the little U.S. on both of my lapels. When I got to the Detachment someone noticed and said that I sure could not meet the Secretary Of The Air Force being "Out Of Uniform". The Detachment staff thought it was funny but someone came up with some U.S. insignia's for me to borrow. Later that day when I returned to the Detachment there was a large poster board on an easel at the front door. An artist on their staff had drawn a cartoon of a First Lieutenant Capps with "US" on one lapel and "THEM" on the other. I carried it back to Olmsted and it was in my office until I figured I had been humiliated enough.

The hardest part of setting up the event was having the Secretary Of The Air Force to swear them in. I had asked the guys in the Pentagon Information Office to handle the details on that one. They had said he agreed to do it after they told him how many press people would be covering it. The Recruiters who had signed up the 18 year old boys (many from small towns in Virginia) had also convinced their hometown newspapers to send a reporter and photographer from each. With the families of all the boys we knew we would have a crowd. The Air Museum especially near the plane they wanted in the background was really not large enough to handle that many people. I worried about a Fire Marshal interrupting our publicity event. Zuckert was on a tight schedule. I was warned ahead of time that there could be no delays. They gave me a precise time the recruits had to be lined up and ready to be sworn in. I was not aware of just how precise they meant.

A limo pulled up right on the dot of the time I had been given. Zuckert walked in with an entourage. He shook hands with the Detachment Commander and I, asked for a script, had the boys raise their right hands, swore them in, posed for one picture with them, wished them all luck and he was whisked away. I could not believe it but one of his assistants actually was looking at a stop watch. One of the photographers who missed the one photo opportunity was asking if one of the others would share theirs. The Detachment Commander, who just happened to be a Lieutenant Colonel handed me a stack of papers. He said, "Capps I'm not sure how you are going to handle this but the Secretary needs to sign all of these". It was the papers for all our new inductees.

I wasn't sure either but when I got back to the Detachment in Arlington I called my friends in the Pentagon and asked. They told me not to worry and come on over to their office. I had been there before but not while missing my little U.S. pens on my uniform. I had given the ones I borrowed back when I came in the door and found my cartoon. I had to go borrow them back. Another Lieutenant from the Detachment wanted to go with me to the Pentagon. He was impressed that I had been there before. We arrived and one of the Information Officers met us at Security. We were on our way to the Office Of The Secretary Of The Air Force. I expected him to not be too pleased to have to sign a stack of enlistment papers. He was not in. He had been hurried off from our event probably to some high level meeting. We met his secretary and the Info Officer told her why we were there. She said "No problem, I can handle it". That's when we saw a machine sign all our papers with a signature which I doubt even a handwriting expert could tell was not his. She said "There are not many of these machines in Washington". Rank does have its privileges.

I said that I worked closely with the Advertising And Publicity NCO's in our seven Detachments. Even though they did not report to me they kept me informed about many of the things they were doing to promote the Air Force. My staff worked to support and help them on many of those efforts. Just as they did not report to me directly I also had indirect contact with the Chief Of Advertising And Publicity for Recruiting Service Headquarters. Those Headquarters were located at Randolph Air Force Base in San Antonio. The Commander was Major General General Neil D. Van Sickle. Colonel Torresson reported to him. My counterpart at the Headquarters level was Colonel Dean Hess. Even though none of us with that title in the seven Groups reported to him he stayed aware of what each of us were doing. We all admired and respected Colonel Hess. He was famous.

Colonel Dean Hess was a pilot and an ordained minister. He served as a combat pilot in France during WWII. He flew a total of 63 combat missions in P-47's. He returned to civilian life after the war but was recalled during the Korean War. He flew a total of 250 combat missions. He was best known for being involved with the "Kiddy Car Airlift", the documented rescue of 950 orphans and 80 orphanage staff from the path of the Chinese Advance. He wrote his autobiography *"Battle Hymn"* in 1956. It had been made into a movie by the same name where he was played by the actor Rock Hudson. Colonel Hess had used the royalties from his book and the film to fund a new orphanage in Seoul, Korea. He and I talked several times about his relationship with the actor when the movie was being made. He did not like Rock Hudson because he kept trying to make it seem there had almost been a romantic relationship between Hess and one of the Nuns from the orphanage. The Colonel said nothing could have been further from the truth. I still have a copy of his book with a personal message he wrote to me. He died at age 97.

AWARD SOME CENTERS OF INFLUENCE

Colonel Torresson asked me to try and come up with something we could do to recognize some COI's from some of the school systems in New Jersey. Several High School Principals and Guidance Counselors from around the state had gone out of their way to assist Recruiters in their area. They helped to steer some young men and women toward the Air Force when they thought either they were not ready for college or could not afford it at the time. Recruiters had sent the Colonel a letter by way of endorsement from the Detachment Commander in Newark. It had a list of people they wished us to acknowledge for their efforts. Being on the Junior Officer Council, I had been involved with a couple of what we called

"Orientation Trips" to other Air Force Bases. The idea was to expose Junior Officers to what our contemporaries at other bases were doing so as to broaden their interest into considering the Air Force as a career. Sounded like a sales pitch to get to go on a boondoggle trip and for most of those base visits it was.

I thought why don't we invite the educators on a trip to San Antonio. I had just gotten back from a trip to Randolph Air Force Base for a meeting I attended with Colonel Hess. I thought that maybe we could possibly borrow the General's plane. He flew in what I guess was the Air Force version of the Learjet. I believed it would seat at least 8 or 10 people. I called Colonel Hess and told him my idea. I had shown him photos of all the displays I had been adding. He could not believe the Hound Dog Missile. He thought the idea had merit and promised to ask the General about use of the plane. I knew that Colonel Torresson had flown on it several times when all the Group Commanders got together. Meanwhile I got on the phone and talked to Information Officers stationed at Randolph, Lackland and Brooks Aero Medical Center (all in San Antonio) to see what kinds of tours could be set up.

Around two weeks later a Captain from our Newark Detachment and I escorted High School Principals and Guidance Counselors on a four day trip which we called The New Jersey Educator Tour. It was a smash hit. Letters of praise were sent by all attendees to the Detachment Commander, Colonel Torresson and the General. They all sent me copies with notes of approval on my pulling it off. One letter which Colonel Torresson received from a young rather attractive Guidance Counselor, I would hear about for many years (and I do mean many). That one I will try to explain.

The Counselor had written in her letter of praise for the trip one paragraph that caught his eye. It said "Particularly, I would like to commend Lt. Capps for the excellent, unceasing concern and care for the group. During the entire four day period, his attention to details, organization, fulfillment of special personal requests made us all feel we were particularly welcome in Texas. I cannot praise him too highly, and I know the other members of the group feel as I do". It was the "fulfillment of personal requests" he wanted me to explain.

JUNIOR OFFICER ORIENTATION TRIPS

My contribution to the Junior Officer Council was when we would have one of our Orientation Trips I would call the Base Information Officer and set up tours of facilities and functions on the bases we were going to visit. At least that made the trips sound legitimate as far as exposing us to other things the Air Force had to offer. I did have a knack for making a case for the reason these trips were a good idea. They sure helped morale. The deal on all our trips was they were at our own expense. No TDY money was used. We would find a pilot and co-pilot needing flying time. The bigger the plane the more people who could go. We went on everything from the old "Gooney Bird (C-47) to a C-54. We had to use up our own leave time. The pilots were required to get their flying time so us tagging along cost the Air Force nothing. In essence, we were getting a "Hop" to and from whichever base we were going to visit. Most of the Info Officers I talked to knew the trips were boondoggles. They would actually have things of interest for us to see and there were always a few officers who were career oriented and wanted to see them. The largest majority were like me and went for the fun.

LABRADOR AND NEWFOUNDLAND

The first trip I went on was to Goose Bay Air Base, Labrador and the Ernest Harmon Air Force Base, Newfoundland. Neither were exactly tourist locations and we knew none of us were ever likely to visit those areas in civilian life, so 16 signed up just for the adventure. We had not chosen either site as high on our preference list but one of the pilots we knew said he and another pilot had to go both places and would be flying a C-54 so they had plenty of room if our group wanted to tag along. When the list went around I think it said, "Here Is Your Chance To Visit Where You Hope Not To Be Stationed". Boy was that true. We arrived at night. It was dark so I think it was night. Walking from the plane to the terminal building we entered a long hall. Someone asked the airman who was leading us if they got much snow. He just shrugged and said, "Touch the walls of the tunnel". That was when we realized it was not a hall but a tunnel with seven feet of snow as walls on either side. Plywood had been placed on the top with a string of lights hanging down the center. Yes they had snow ….or rather it had turned to ice. We were taken to the Visiting Officer Quarters. With so many of us, many had to go three to a room. Bunks were all we needed so no one cared. We all headed to the Officer's Club. On both bases it was the center of activity. We had flown in to Goose Bay first and spent one night there. We all did take a tour of the base as there was not much else to see or do. The "O" Club was First Class. We were only supposed to be in each place a day and night each but our trip was on the 17th of November and the weather was unpredictable.

Arriving in Harmon, it was snowing pretty hard. In fact by the time we checked into our rooms one of the pilots said, "Might as well make yourselves comfortable. We are not likely to get out of here tomorrow". We ended up snowed in for three days.

Some of the guys complained because we were all using up our leave time and at that moment we did not know how long we would be stuck there. I made the best of it. The Officer's Club there had slot machines and I loved to play.

We had gone to their Base Exchange in the afternoon. I had never seen an exchange stocked like that one was. I guess because those stationed there needed things to keep them occupied during bad weather. It had a whole section of plastic models, cars, boats, and airplanes which you could build. There was another section of puzzles and another full of art supplies (everything was very inexpensive). I bought up a supply of art stuff and of course my allotted gallon of booze. That night I hit a Jackpot on a quarter slot machine and won
$500.00 Back then the quarters fell into a tray making a lot of noise. You can imagine 2,000 of them falling attracted the attention of most of the guys in our group. I kept filling up cups as the tray overflowed. I more than paid for my trip, a few bar bills for my friends and had money left over.

On our second night at the club there were two guys from the Seagram's Distillery. They were showing a film in the bar (old 16 mm projector) showing how Seagram's was made. They announced that as long as the film was running they were serving Crown Royal for 25 cents a shot and there was Seagram's Seven Punch available for Free. Guess which one all the Junior Officers went for? When the film was over everyone started shouting show it again. They did … several more times and the drunker everybody got. Finally the two guys themselves started drinking too.. I think by the time all their booze ran out we even watched that film backward. Good thing we were snowed in. Our two pilots were in no shape to walk much less fly.

I met a civilian girl at the club who was on an assignment, sort of like a military tour, where she was spending two years teaching there. When I got out of service and got back to Birmingham, I ran into a girl from my high school class, I found out that she had been in Harmon at that exact same time as a teacher. Funny that I did not run into her. Small world. We finally made it back to Olmsted with stories to tell of being snowed in.

THE FIVE SIDED PUZZLE PALACE

Now that I was a fully qualified Information Officer, I had contacts with every Air Force installation and the Pentagon. Over a period of time I helped set up quite a number of different trips. Since we had a Helicopter Squadron stationed at Olmsted I set up a trip for a small group to fly by chopper to visit the Pentagon, which everyone called the Puzzle Palace. Interestingly enough the helicopter pad where we landed was near the entrance where the American Airlines Flight 77 crashed into the Pentagon 37 years later on 911. The Information Officer who took us on a tour could only take us on a small one because so many parts of that famous five sided building are in classified areas. He did let me stop by and say hello to my old one time boss General Favor. On one of my several trips to the Pentagon another First Lieutenant and I had just left a visit with Brigadier General Favor. We were heading down on a long escalator when we realized that in front of us all we saw were stars … many stars ... on shoulders and stripes on sleeves. A meeting of the Joint Chiefs had just adjourned and they and all who had attended the meeting were on the escalator the same time we were. Maybe we were acting nervous or possibly one of us pointed but all at once a full Colonel behind me tapped me on the shoulder and said, "Don't worry guys in this place they have forgotten First Lieutenants even exist". I'm not sure who all else was on that escalator but I remember Air Force Four Star General Curtis LeMay, Army Four Star General Earl Wheeler and Navy Admiral Thomas Moorer.

EGLIN AIR FORCE BASE

Probably the best trip we ever made was to Eglin Air Force Base in Florida. When I called the base Information Officer at Eglin and told him what I was working on, he laughed and said, "Let me tell you when to come down here. I can guarantee you will fill your roster within three minutes after you announce it. We are in the process of setting up a firepower Demonstration for several high ranking Government Officials, Dignitaries, Pentagon Officers, Congressmen and Senators. We are having a grandstand set up and I can get your group seats to attend the event. Talk about a trip of a lifetime.

First I started calling pilots (I had a large list). I asked who needed flying time and if we could get a big plane. The first pilot I called was a Major who said he did not need flying time but said for something like that I'm going. He called me back in five minutes with two pilots and two more like him who just wanted to see the demonstration. He had also lined up a C-54 which had the most seats of any on base. I called my friends first and put them on the list … then it was first come first served. All phones in my office started ringing. Word spread fast. Even a Lieutenant Colonel called to say the Lieutenant in his office could come as long as he too were included. It was filled with fourteen on the waiting list hoping for cancellations.

I am afraid that the plane was probably overloaded but we took off on schedule. One of the first Junior Officers who got on the list was our Base Ophthalmologist who had never gone on any of our trips. In fact he had never flown on a plane in his life. Believe it or not he had been in the Air Force almost two years and had never been on a plane, much less flown on one. I don't remember if we had a scheduled refueling stop planned but when we got down near South Carolina, the pilot came on the speaker to announce that we had lost an engine and would be making an Emergency Landing at Shaw Air Force Base. Our Eye Doctor was more than a white knuckle flier. He had not said a word to anyone on the flight and the Lieutenant sitting next to him had gotten up several times asking if anyone could maybe say something to his seatmate to calm him down and make him relax.

On hearing the announcement, he went into a panic and started crying and saying over and over "We are going to die, we are going to die". The Major and the Lieutenant Colonel, both Command Pilots, assured him that we were fine. We still had three engines and were in no danger. It was just procedure to declare an emergency so we could be cleared for immediate landing. He calmed down a little but when we started coming in he saw Fire Trucks and an Ambulance and other Emergency Vehicles running parallel to the plane as we came in. He turned white as a sheet.

The landing was smooth and we taxied up without a problem. Everyone was laughing and cutting up.

Someone even said, “We almost had our Firepower Demonstration early”. The Eye Doctor did not laugh. After about thirty minutes or so hanging around Base Operations, it was announced that we would have another plane which would arrive soon and take us on to Eglin. When it came time to reload I was checking off my list to make sure everyone found their bags I could not find the good Doctor. We found out that he left in a cab and told someone he was taking a bus back to Olmsted. As it turned out our group went on to Eglin, saw the demonstration, toured the base, spent two nights and went back to Olmsted. We beat the Eye Doctor back by a day.

The demonstration was unreal. We saw Napalm and how scary it must have been if you were on the ground in Viet Nam. They also showed us the tracer bullets fired from a circling C-47 which was named “Puff the Magic Dragon”. We saw low flying C-119’s come in like they were landing. The rear door would open and a large parachute would drag out jeeps and trucks but the planes never actually landed. Parachute jumpers were landing with black chutes. Not only were they hard to see coming down, we could barely hear their planes. Impact bombs were dropped in a field nowhere near the grandstand but we felt the shock wave when it came and hit our faces. It was all a surreal experience. It convinced me that if I had to go to war I wanted to be in the air.

BERMUDA

Another Orientation trip was to Bermuda. Our base there was not a large one but we did have one named Kindley AFB. The Navy had the biggest presence on the island. The plane was another C-54 and it was set up to carry quite a few passengers. Word always got out fast when we planned one of these trips and the list filled up within minutes. There was always a waiting list because there were last minute cancellations when people found out for one reason or another they could not take leave. Like the trip to Eglin we had a long waiting list. I do not remember how many made this trip but I do know we had a full load.

When we landed at the base the Information Officer was waiting on the tarmac. He got on the plane and made an announcement. "Lieutenant Capps has informed me that all of you made this trip because of your desire to learn more about the Air Force and other bases. Outside you will see a bus for all of you who wish to go straight to the Visiting Officer Quarters, the Officer's Club or bike rental place. You will also see my Volkswagen Bug. It is for all of you who wish to tour the base". To loud applause, he and I left in the VW and everyone else boarded the bus. Of course he dropped me off at the BOQ and said call me if you need me.

Bermuda was great. Many of us rented mopeds and toured the island trying to get used to driving on the opposite side of the road. We did fine until we reached a curve (which was often). Eventually we broke into small groups of two to three as most of us differed as to what was of interest. Two of us decided to ride over to see the Navy Base as they had a larger contingent on the island than the Air Force. At the Main Gate (since we were not in uniform) we showed our ID cards. Lieutenants in the Navy get more attention than the Air Force so we were treated pretty nice and even had an escort leading us to their Officer's Club. We had lunch there and headed back toward our base. It was getting late and we needed to decide if we wanted to turn the mopeds in or keep them for another day. We went straight to the rental place.

As we pulled up two girls caught our attention, both pretty cute. They were debating about renting bikes but the instructions about driving on the left had them worried. We introduced ourselves and suggested that maybe it would be a good idea to just rent helmets and join us on the back of our bikes (which we had quickly decided to keep for another day). They said they did not know about just going off with strangers. We went into our sales pitch mode fast, explaining how we were 'Officer's And Gentlemen by Act of Congress", what could possibly be holding them back. No telling how many times that line had been used.

After introductions we found out the were Airline Stews, they rented helmets and off we went.

The one named Michelle ended up on the back of my bike. I am not sure how the choice was made but it seemed to be accepted by each as to who went with whom. Michelle was a brunette and Gloria was a blond. We started off by driving them to spots we had already been …... acting like regular tour guides. Of course it was now getting late in the afternoon and we would soon be running out of light. After three or four stops to take scenic photos, Michelle whispered in my ear that they had both been there many times and probably had seen every part of the island and knew more about where to go than we did. They suggested a bar which was built in a cave. We ended up there just as the sun had begun to set. After drinks my buddy and Gloria said they were heading back to the Officer's Club to have dinner. Michelle suggested that the two of us just grab something in the bar. It was after 11:00 before we left. Michelle had asked if I played golf. At the time I had not but I said it was a game I wanted to learn. I knew very little about the game and even less about the jargon which went with it, It was obvious that she had played a lot. Her dad belonged to a club and he had bought her lessons for her birthday. She threw out a couple of Golf Pro names I had heard of who she said she met at a tournament.

Driving back toward the hotel where she was staying she told me to turn on a drive to a nearby resort. She wanted to show me a golf course which she had played. We left the bike and climbed a fence which led to the course. Walking up a steep hill we were soon overlooking most of the course and the ocean in the distance. There were three sets of tees with a par three anywhere from 180 to 150 yards to a green down below. She explained how depending on your handicap which tee you would choose. She told me what a greenie was, a birdie, a bogey and par. I said I thought I would like a game where if you were not up to par ...it was considered good. She laughed and said, Oh you will learn all the terms". She went on to say that she rarely sliced, she was mostly a hooker. Who knew?

THE PAD IS BORN

My little apartment, now that Jake and I were sharing it had gotten much smaller. We both traveled a lot but when we were both in town at the same time we were tripping over each other. We decided to look for something bigger. Our office next door neighbors, Charles Earnhart and Del Fisher (the OSI Agents) met us for lunch quite often. One day we were talking about moving. They both lived in the BOQ but were now eligible to live off base. They suggested the four of us try and find a large house we could rent.

When Jake moved in I worried about my landlady not approving but we gave her $80 instead of $60 without her asking. She was happy and I was saving $20 a month. When we moved out a pilot on my list was looking for a place. His wife and kids had moved to Atlanta for her to care for her mom and he needed to move out of base housing. We told him the rent was $100 a month so our landlady was more than happy when we moved.
Jake and I had looked at houses before the four of us decided to try and find a large place. We had been shown two which were total disasters. A pushy woman from a rental agency had helped us. The first one she took us to happened to be in Harrisburg. Which was an okay distance but we preferred to be closer to the base. The woman had said this will be perfect for you two, it's completely furnished and it has a lot of things you can keep. Which seemed to be a strange selling point. She said,"The owner is offering a deal".

That should have been our first clue. When we walked in, the front room was full of boxes, stacks of old newspapers and magazines. And you would not believe what all else. The woman who lived there had died and she was obviously a hoarder. Her son wanted to rent it as is and the deal was, whoever cleaned it up and disposed of all the clutter could live there rent free for two months. After that it was $125 a month but that would include water and electricity. When we found out she had died there we did not look further than that front room. The second one the same rental agency lady showed us was in Highspire which was close to the base. When we walked in it there was the smell of smoke. When I asked about the smell she said there had been a small fire in the attic but it had been taken care of and the smell would probably go away. I think Jake asked how long ago the fire was but she suddenly did not know. It smelled like the end of a bonfire at summer camp. Since we did not plan to sit around singing Kumbaya we passed on it.

I was not the only Lieutenant who found a moonlighting way to make extra money. One Lieutenant in his spare time sold house trailers These were not like Airstream Travel trailers but the kind you saw up on blocks for more permanent living quarters. The Lieutenant was from Panama City, Florida. I had only known one person from there, a girl I dated at Auburn. When I asked if he knew her, thinking it was probably a silly question because Panama City was not a small town. He said that I would not believe it, but they grew up as kids as next door neighbors. Again Small World. At any rate the four of us Lieutenants talked to him. They not only sold trailers they had a trailer park where they rented some. We thought maybe we could rent two next to each other.

He came up with a better offer. His boss owned a three bedroom, one and a half bath house which had been vacant for a good while. It set back from the road on two acres and he needed to rent it to someone who could take care of the house and the big plot of grass. It would come with a riding lawn mower. It was not furnished but he had a warehouse full of furniture which he used to furnish trailers. We could pick out whatever we wanted (and loan us a truck). We all went to see it. Three bedrooms instead of four posed a problem until I saw a finished attic space which was the width of the house. It only had a small window at each end, was somewhat dark but I could visualize setting up my bed at one end and an art studio at the other. I volunteered to take it as my room.

I doubt any four people could have been more compatible. We had more fun in that house. We took turns cutting the grass. If fact it was fun riding that mower so we often flipped a coin because no one minded doing it. Our address was actually Hershey, Pa. (yep, where they make the candy). We were a long way from the Hershey factory and the Amusement Park as we were right on the line between Middletown and Hershey. At least a half of the two acres was over the Middletown line. We were within walking distance of the Susquehanna River and in the winter with no leaves we could almost see Three Mile Island. The house had a nice kitchen, a small dining room and a large living room with a working fireplace.

We went thru two cords of firewood. Whoever got home first always started a fire. The ashes could be dropped down a trap door in the fireplace to the basement where there was a large metal container to hold them until they could be disposed of. On the side of the house there was a strawberry patch which came back every year. We had so many strawberries that we gave them away to neighbors. At first we had a problem with turtles from the river eating them until we added a low fence. A policeman's wife who lived across the street would trade us strawberry pies for a share of our crop.

The number of parties we had at the pad has been long forgotten but our place was well known by other envious Junior Officers on base. For our one Christmas we had a real tree. Since none of us had decorations we thought why not have a party and everyone bring something to hang on the tree. To make it special we said you cannot bring traditional ornaments. Bring something unusual which you would never see hung on a Christmas Tree. I bought different colors of Day Glow paint and a Black Light which was an extra we had for my Space Capsule display. We painted each item our guests brought. As soon as it was dry they hung it on our tree. No way that I can remember what all we ended up with but some of the more memorable ones were a beer can opener (better known as a Church Key), a salt shaker swiped from the Officer's Club, A rat trap, a fly swatter, and a bra where a star usually goes. Everything could not be painted with the Day Glow because it would not stick, such as several different brands of beer cans. Someone brought a garland made with hundreds of paper clips. It was a lovely tree.

I said we could walk to the Susquehanna river. An African American NCO who was in my same office building and I went fishing around three days a week when we were both in town. You couldn't catch much other than Carp but they were huge and fun to catch on a rod and reel. We did not keep them and had been throwing them back when the Sergeant made friends with a fisherman we saw there a lot. He kept his and said if you knew how to clean them and prepare them they were good to eat. The best bait was a dough ball but we had trouble making them stay on the hook. The sergeant cut a deal with that other fisherman ….he would supply us the bait and we would give him all we caught. He not only knew the recipe for preparing the Carp, but he had a secret for making the dough balls. He would never tell us the secret to either.

My fishing buddy ran our Motor Pool and with seven Detachments in as many states with Recruiting offices all over those states (all with vehicles) ….it was no easy job. He also was an Expert Marksman with a 45 and won many trophies in Air Force competition. At night he had access to the Base Firing Range and many nights after fishing we would end up there. He tried but could never get me to qualify with the 45.

We had the house over a year. When we moved in we all agreed on House Rules. They were never written down but we never broke any of them. Your room was a castle. Do anything as far as leaving clothes scattered around or any clutter you wanted. The rest of the house was to stay immaculate. Pick up after yourself. No dirty dishes were left in the sink. The kitchen stayed clean. If anyone saw the garbage can full they took it out. In other words, other than each person's own bedroom the house was ready for visitors all the time. No smoking inside. No cigarettes, cigars or pipes. This was also for any guests. The only smoke we wanted to smell came from the fireplace. Jake and I were the only two who were gone a lot. Charles and Dell worked as Investigators on the base and around the area. They were home each night. Charles and I were handy with tools. We repaired many things when we moved in and the owner of the house was aware of all that we did. He knew the shape the house had gotten into while it stayed vacant so long. He loved us as renters. I was his main contact and always took him the rent checks.

I guess all good things eventually must end. Dell got orders sending him to Korea. We replaced him with someone the rest of us knew well ….the Doctor who was the Base Dentist. He also loved to fish so he joined the Sergeant and I at the river. When he was "On Call" he carried a two way radio with him when we went to fish. We noticed when he would test it, as soon as we arrived, other fishermen would pack up and leave in a hurry. Our new buddy who was supplying us our bait finally told us why. None of them had fishing licenses and they thought he was the Game Warden. It opened up some of the best fishing spots. So we never told anyone that he was just a Doctor "On Call".

Our new four took up where our original left off before Dell left. Then Jake decided to get married. He had kept writing and calling the girl he left behind in Amarillo. I knew her well because I had double dated with she and Jake. She was beautiful and we all all wished them well. His replacement was a mistake. This is an understatement. He broke all unwritten rules from the start. He also broke laws by bringing underage girls to the house. He was more interested in notches on his belt than joining our unofficial fraternity. The final straw came when Charles got orders to go to Viet Nam. I told the Doc that after Charles left I would be the Last of the Mohicans and I had had enough of our last replacement so I would be moving out. He being a Doctor at the only Clinic on base (and the fact there were no HIPAA Rules back then) told me he was planning on moving out also because our last new housemate had contacted VD from one of his frequent visitors. The Doc also had had enough.

I went to give our landlord notice. I told him I had no idea what our soon to be Lone Ranger wanted to do but two of us were moving out. He said that it was a funny thing that he was about to give us notice that he had a buyer for the house. Middletown had finally given him permission to split off the acre which was across the line and he could divide it into two lots. The potential buyer of the house was happy to only have an acre to keep up. Problem solved ….only now I needed a place to move to.

Since the owner of the house also owned the trailer sales lot I asked him if he had a suggestion for me. He said, “As a matter of fact I do. You remember that I own a Trailer Park in Highspire. I need someone to act as Manager and be on the property. There is a small vacant trailer right at the back entrance which I will rent to you cheap if you will be my contact at the Park”. We left immediately to go look and see the trailer. He was right. Compared to the trailers who were renting space in his lot, this one was small. It was about 60 feet long, not very wide and still on wheels, where most of his rentals were up on blocks. It was connected to a septic tank and had running water. The set up was simple. The left side had two doors, one at the front and one at the rear. You walked in the front door and to your right there was a tight living area. Next was a small galley. This led to a middle bedroom, then thru a sliding door into a bath with tub, commode and sink. It was so small you could sit on the toilet and shave using the sink directly in front. This led thru another sliding door to the back bedroom. At my first look with no furniture it looked fairly big.

His offer was too good to pass up. I would be the contact for all residents. If they had a problem, I would not have to handle them ….just bring them to him. I would not collect rent. They all had to mail it directly to him. Sounded like I had little to do as Manager, other than be his contact. All utilities (which were only water and electricity) he would include for $60 rent a month. He would also install a phone which he would also cover.. Sounded great so I accepted on the spot. He had one of his employees help me find furniture and move it from his warehouse. I set up the middle bedroom as an art studio. All of a sudden it did not look as big. We moved my bed from the house which made the back bedroom wall to wall bed. That made the back door only usable as an escape if I had a fire. I made it do for a long time ….. all the way until I got married.

LIFE AS A TRAILER PARK MANAGER

It was one of those experiences you only want once in your life. First just living alone in a trailer made me realize that I never wanted to do that again either. As the "Manager" I would hear complaints, none of which I could handle...only pass them on. I was awaken by the Police twice when a wife three trailers over locked her husband out in the snow wearing only his underwear. They thought as the name "Manager", it was my duty to let him stay warm in my "Office" until they convinced his wife to let him back in. As I have said before You Can't Make This Stuff Up. And yes it did happen twice. The other really big thing happened when I was out of town. In the middle of the night someone had actually stolen a trailer. None of the neighbors cared for the people who lived in it so nobody reported it being missing. I only discovered it after being gone two days. When I called my trailer court boss he said they owed him back rent and it was his trailer and not theirs. With such a head start it took Police in Arkansas two weeks to find it. I have no idea what the result was. I only remember the owner saying that I should have heard them moving it. It was located near the main entrance and I was by the back one, also I never told him that when it happened, I was out of town. Since my trailer was on the end by the back exit it was next to a house on an adjoining street. A sweet old lady lived there alone. I had a small patch of grass along the side in front of my two doors. She had a push mower in back of her house. I borrowed it and cut not only my small yard but hers as well. This became an every two week project. She had told her son who stopped by one day to thank me and say he appreciated me keeping an eye on her house and doing some of the chores he usually had to do. She was nice and loved to talk when she was in her yard planting flowers especially when she saw me coming or going. She said that she bet I could write a book about some of the stuff that went on in the trailer park. She even suggested a name "Peyton Place On Blocks". Since I was gone so much I saw very little but every now and then I did hear gossip about things which happened. So in so is having an affair, Mr. Franks gets drunk, the woman in the blue trailer entertains different men, that sort of stuff. My neighbor in the adjoining house would occasionally bring me a pie on her baking days.

FEMALES ON BASE

There were lots of women working on base. In our two buildings we had several WAF Officers and NCO's plus each Division had a civilian secretary. I never understood why, but the secretaries were hired and supplied to us by the Base Personnel office. We were not given an opportunity to interview, approve or disapprove of who they sent.

I was lucky. I inherited a good one from the Captain I replaced. She was so good that Colonel Torresson had her moved over to become his secretary. He explained it as being because he shared his with our Chief of Personnel in our Group so he needed the best typist.

Base Personnel sent me a Doozy or maybe I should say Floozy as a replacement. You will see what I mean. She could type but not great. Her main problem…. she would always be late or never show up on Mondays. I was not there every Monday so it was my staff who brought the regularity of it happening to my attention. On one Monday when she actually showed up on time she had a copy of the Harrisburg newspaper. On the front page was a photo of a raid which had taken place downtown where a group of prostitutes had been arrested. She was laughing and showing everyone the photo and calling the "Ladies" by name. She seemed to know them all. We thought this to be a bit out of the ordinary but since she had grown up locally maybe she just knew them from high school. The next Monday stuff got worse. I had warned her it had to stop. I was told that it was really difficult to fire a civilian and you had to build a case in order to do it. I had a file which was beginning to get thick.

One Monday at 10:00 a.m. she had not shown up. She called around 10:30. My First Sergeant had answered the phone. He said, “It’s her, you better take this”. She sounded a bit groggy but said “I’m sorry but I just woke up”. My reply was “Well it is only 10:30 you can still come to work”. The straw which broke the camel’s back was when she replied “I know, but I woke up in Philadelphia”. I told her that when she returned to Harrisburg to stop by and clean out her desk. No one questioned my firing her.

Two weeks later she made the front page herself. Her boyfriend (or maybe her pimp) who used to bring her to work and pick her up in his Jag convertible had been shot by her. He was not killed because she shot him in the buttocks as he was climbing out a window. She caught him in bed with another woman. No, you can’t make this stuff up.

One female, who I worked with a lot was a WAF Lieutenant who worked with Jake in our Education Division. Her name was Mary Abbott. She was from Mississippi and had the greatest Southern Drawl. The Officers who worked in that Division often traveled to colleges and universities to speak to students about making the Air Force a career as an officer after graduation. I would tag along on many of our Education Staff’s trips to different colleges. I would set up publicity with many of them including photo shoots and occasionally radio and TV. My favorite trip with Mary was to an all female Catholic College. The Nuns had all the Seniors seated in their auditorium. Mary gave her usual presentation and I think showed some slides of various bases many of which were in scenic beach locations. At the end she took questions from the students. One girl stood up and asked "With so many men, how do us girls keep them from always hitting on us"? Without hesitation Mary replied " My best advice is…. Keep your legs together, your skirt pulled down and walk home in a crowd". Even the Nuns could not keep from laughing.

PENN STATE

One of my favorite Universities to visit with our Education Division Officers was Penn State. I probably went there at least four times. We would set up a display in the Union Building. On one of the trips Coach Joe Paterno stopped by to ask why we were on campus. When we told him he just said "Don't recruit any of my football players". We told him we were only interested in them after graduation. He seemed pleased with that answer and invited himself to join us for lunch. He was only the Assistant Head Coach at the time.

Another visit we ended up accompanied by an FBI Agent, who had an office in State College. During an earlier visit, I had placed a large road sign which said GO AIR FORCE with reflective letters on the road going into the campus. It had been stolen. As Government Property, the theft was a Federal Offense. The Agent met me at the Union Building. He told me he had already located my sign. He then added "You might want to just leave it where it is. I believe it will get more attention where it is, than on the road". It was hanging in the party room of a well known Fraternity House. I left it.

The Agent assisted us again on another visit. I had secured a Wind Tunnel Model of one of our jets to display in our booth. It was a beautiful scale model and valued at $5,000. We had gone to lunch and it was gone when we returned. The Agent said that it was most likely a Fraternity Prank. He went on the Campus Radio Station, told what the model was worth and the jail time which might accompany such a theft. He added that if it was back within 15 minutes no questions would be asked. It was back in ten.

The visit to Penn State I remember the most was with a Captain from our Education Division. He was also one of our pilots in Group Headquarters. He was the epitome of the Air Force Poster. His shoes were so spit shined you could see your face. He always wore his Dress Blue uniform with his wings and all his ribbons. He was Hollywood Handsome and our most photographed officer. He always gave the same speech and it was usually to guys in a fraternity. He was so good that word got around to different colleges that he was requested to come and speak. I heard it so many times that I had it memorized. I was with him at Penn State. The frat house party room was packed because they had invited friends (all guys) from other fraternities.

He would start out "There I was….. 5,280 feet up…….flat on my back (long pause)
In a bed, in a hotel room in Denver, Colorado. What brought me to this hotel room? (another long pause) Other hotel rooms all over the world. I looked up at her and said "What makes your lips so moist"? She sniffed, wiped her nose which was running and said "I don't know'. By now everyone is laughing. He then said "I am a pilot in the United States Air Force. You will ask. Why is that so special? ….I am going to tell you. How many of you can say you peed on Ohio State from 10,000 feet'? Wild applause because he always picked the biggest football rival of the college where he was speaking. He went on " So what do I make as a pilot in the World's Greatest Air Force"? He would then quote his salary which included his flight pay, housing allowance and he would always throw in that he was being paid extra (TDY) for speaking to them. He said, "Okay if you are still single what will that amount of money buy you each month"? He broke it down to number of dates you could have, how many bottles of Jack Daniels, how

many steak dinners, a subscription to Playboy and a gross of condoms."And believe me guys this uniform is a chick magnet. Sometimes he would even throw in "Look at Capps over there even he got a date here at Penn State". I never saw him not get a standing ovation. He was a perfect Recruiter for college graduates. I flew with him in the U3A. He was also a damn good pilot.

SPEAKING OF PILOTS

I have said before that Colonel Torresson was a Command Pilot. That was one who had at least 3,000 hours in the air and not less than 15 years flying service. He had had much more than that. At the beginning of WWII he was a flight instructor with the Army Air Corps at West Point and Stewart Air Force base in Newburgh, NY. He had one of the earliest date of ranks of any officer at Olmsted. That's why he had one of the nicest quarters right next to the Officer's Club. I flew with him many times.

One time we were going to our Detachment Headquarters in Louisville, Ky. The cross wind that day was unbelievable. The tower had advised that it was dangerous. I had witnessed many cross wind landings when I was stationed in Amarillo. The wind was always blowing there. Sometimes it looked like the planes were coming in sideways.

Watching a cross wind landing is one thing, being a passenger is completely another.

I was the lone passenger in the back seat of our U3A (Twin Cessna). The Colonel used the Crab method. There are two methods . The Wing Low and the Crab. On final approach he was crabbing into the wind to prevent drifting left or right of center line. He maintained the crab to just before touchdown then stepped on the rudder to align the nose with the runway. He used the ailerons to prevent drifting. We touched down as smooth as silk. When we got to base ops there were two Navy Pilots who had landed earlier that day in two F4C Navy Jets. They applauded the Colonel when he walked in saying that it was the best cross wind landing they had ever seen. We were still in Base Ops when they took off. The Operations Chief told the Colonel to walk outside that the two Navy Pilots were going to offer him a Salute. When they took off side by side they suddenly went straight up and popped a simultaneous sonic boom.

Lieutenant Glick and I saw another example when we flew with the Colonel to Randolph Air Force Base in San Antonio. We were going to a Recruiting Training Meeting. Howard and I were leaving the visiting Officer Quarters and meeting the Colonel for breakfast. We were walking to the Officer's Club. When we got to a corner we saw a vehicle approaching flying a General's Star from the front fenders. We both snapped to attention and saluted as it passed. It stopped about half block away and started backing up. We had no idea why. The door opened and Brigadier General Hogan invited up to hop in. He asked “Where are you two Lieutenants off to so bright and early”? We said that we were meeting our boss Colonel Torresson for breakfast. He said “You mean Tom? He is the best damn pilot in the Air Force. He taught me how to fly at West Point. I think I’ll surprise him and join you guys for breakfast”. He did and it was a fun time. They talked about the old days and told a few war stories.

After meeting the General, we wondered why Colonel Torresson had not received a star. Over the years I heard several opinions. Some said it was because being an instructor he was never sent into combat during WWII or Korea. Personally I think it went back to Selma, Alabama. They were having racial problems there years before the Selma to Montgomery March of 1965 (and the famous Edmond Pettus Bridge). The Colonel had been Deputy Base Commander at Craig Air Force Base in Selma. In downtown Selma there had been trouble with sit-ins and demonstrations going way back to early 1963. Bobby Kennedy, who was Attorney General flew into Craig. He wanted to set up at the base so he could monitor what all was happening in Selma. The Colonel, afraid those tensions would spill over to the Air Force Base, where there was no racial problems refused to let him use the base for his Command Post. Just my personal opinion but I think Kennedy blocked that promotion from ever happening. Kennedy was a vindictive person. Colonel Torresson would have been a great General.

CLOSE CALLS IN THE U3A

I flew many hours as a passenger in the U3A and one time I got to sit in the Co-Pilot seat. I can only remember two close calls where I got extremely nervous and on one occasion scared out of my wits. We were flying back from a trip to Charleston West, Va. into Harrisburg. Almost in our final approach we lost the right engine. The Major who was flying and the Captain who was Co-Pilot either decided or accidentally feathered the left one. We were landing dead stick. They thought we had enough altitude but coming in with the runway at Olmsted parallel to the Susquehanna River we barely cleared the last bridge before the base. Landing with only the noise of the wind reminded me of landing in the Glider in Israel but this time I was a little more scared. Landing was fine and the rescue squad followed us to a stop. Exciting but I did not want to do it again.

The second incident happened to be with the same pilot (different co-pilot, the Captain who was the good speaker from the Education Division). We were on final approach into Philadelphia when a sky diver almost clipped our wing. He was so close I could see the panic on his face. He was in free fall so his chute had not opened. We could have all been killed. The Major got on the radio and was screaming, reporting the near miss. Air Traffic Controllers had seen it, but too late to stop the plane from dropping the sky divers. It was way off course from their drop zone. I imagine a great deal of paperwork followed that episode.

THE RECRUITER WHO ALWAYS BEAT HIS QUOTA

We had a Recruiter in Harland, Ky. Who not only beat his quota for recruits every month he often got enough extras to help the Detachment meet theirs. Colonel Toresson sent me down to find out how he did it. All of the officers at Group Headquarters took turns inspecting individual Recruiting Offices. Usually it was an office close to where you were going for some other reason. In the case of Harland it wasn't close to anything. There was one road into town and the same one going out. They claimed the movie "Thunder Road" was filmed there. I think I found out why.

When I arrived at the office in downtown Harland (which was on the main street) there was a sign on the door which read "CLOSED … Lieutenant turn your radio on to station 640. I am at the Little

League Park just down four blocks". I turned on my radio and immediately an Air Force Spot came on. It was one that I had gotten made by an Ad & Publicity NCO who moonlighted as a disc jockey. I started looking around and across the street on a second floor was a radio station. A bunch of people were looking out the window waving. He had set it up for them to play the spot as soon as they saw my car.

I drove to the ball park took off my jacket and tie and joined him practicing baseball with the boys. After practice we went back to his office. I did the Inspection thing which amounted to checking off stuff on a printed sheet. Then we the got down to business. "How do you recruit so many kids from a little town like Harland". He said, " Lieutenant follow me". We walked around the block to the Sheriff's Office. He introduced me to the Sheriff. He told the Sheriff my question. He replied, "Simple, Lieutenant, kids get in trouble around here quite often. I just ask them which they prefer . …..serve some time or serve your country? They usually volunteer to go speak to Sergeant Hatfield. Solves my problem and his". He then asked who my boss was. When I said Colonel Torresson he said, "Well I want you to take him this. Tell him it has been tested in the lab and it is safe to drink. Best damn moonshine in the whole state". With that he handed me a quart fruit jar filled with pure white lightning. I think he said it was 170 percent proof. All I remember is that the Colonel kept it in his office and occasionally served a drink to visitors along with my office visit story. It was strong stuff.

I got a good publicity story which came from that office. Our recruiter drove an Air Force vehicle which had a Recruiting Seal printed on each front door. One day a bullet was shot thru the seal on the drivers side because a moonshiner thought it was a revenuer's car. Luckily no one got hit. It made a good story for the papers but I always wanted Sergeant Hatfield to recruit a McCoy because it would have made a better one.

THINK FAST AND REVERSE A JOKE

It was my first visit to our Newark Detachment Headquarters. On my second week on the job I had been to Newark but not to our office so I was not exactly sure where it was located. The directions I was given showed it to be in downtown Newark. I had been told to drive to a certain side street and there would be a fenced parking lot with several government vehicles and I could not miss it. There would be an NCO waiting to unlock the gate and tell me where to park. I found it with no problem. The sergeant was there and he said he would be pulling out and for me to park in his spot. He handed me a chain and a lock and said, "Just lock the gate after you park”. He gave me walking directions to their office.

I should have been suspicious when he asked “Aren’t you the Lieutenant from Alabama”? The directions were something like …. go around the corner, two blocks to the right. You will see a blue awning, a restaurant then it will be the second red door just past it. Turns out the Detachment Headquarters was across the street from where I was sent and I found out later that most of the staff were watching as I walked into an office for the NAACP.

I caught on quick, figuring I had been set up and decided to have a little fun with their attempt at humor. This was not long after the Selma to Montgomery march in Alabama and it had dominated the news so I am sure they thought I would not know what to do walking into an NAACP office. I introduced myself and asked if it would be possible to talk to someone in charge. It just happened that the Executive Director for the National office, was visiting that day. I told him I was the Chief of Advertising and Publicity for the Air Force Recruiting Group which covered the area and I was doing an article to send to the Air Force Times and other publications about non discrimination in the Air Force and opportunities for young blacks. I asked if I could quote him in my article. I can’t remember exactly what he said but he was pleased to be quoted. I probably stayed at least 45 minutes but by the time I actually showed up at our Detachment I had made enough notes to show our guys that they had done me a favor. Thinking fast often pays off. It certainly made an impression on the NCO at the Detachment who handled Advertising and Publicity. My first visit to our Newark office was a success but with a weird ending. I had stayed overnight in a hotel and when I went to get my vehicle from their fenced lot the next morning, it was up on blocks with all four tires missing. I had to take one of the Detachment cars to drive back to Harrisburg.

RECRUITERS GET IN TROUBLE TOO

The vast majority of Recruiters that I met were what I considered the best representatives for the Air Force. Occasionally one would get into trouble. In this case it happened to be two. At an office in Maryland we had two recruiters who received serious charges against them. It seems that they been recruiting nurses to go into the officer program and they were doing what they called pre-induction physicals in the back room of their office. It got them kicked out of service and I never heard what happened with the civilian charges. I don’t remember a chapter in the UCMJ (Uniform Code of Military Justice) which covered such an incident (or "uncovered" as the case may have been).

THE COLONEL HAS TWO DAUGHTERS

I had met Colonel Torresson's son Skip early on and we had become friends. In fact we double dated a few times and Skip and I had gone club hopping occasionally. He was in the Air National Guard and I enjoyed hearing what all they did and where they traveled. I knew Skip had a younger sister named Susie. What I did not know was he had another sister named Mary Ellen. One night there was a party at the Officer's Club. Skip was there. He had seen me dancing with a girl and I was trying to show her how to do the old Panama City Bop ….. and with little success. Skip having lived in Alabama twice while his dad was stationed at Maxwell in Montgomery and Craig in Selma was familiar with the PC Bop. He said, "You ought to invite my sister over because she knows how to dance the way you do". I thought dancing with his little sister Susie would probably look pretty funny since she was around 13 years old and I told him so. He said, "I'm talking about Mary Ellen, my other sister". Remember the Colonel's quarters were right next door to the Officer's Club. He went to get his other sister. Maybe it was not love at first sight but it would qualify as a close second. She was better looking than anyone I had seen since moving to Pennsylvania and her personality grabbed my attention right away. Why had I not seen her before that time? I would soon find out. She was married and had a son named Mark and had not been living there. We danced and had a good time. She finally told me before she left that she was in the middle of a divorce and had come to stay with her parents. I asked but was told that she was sorry that she could not date until the divorce was final. I remember thinking …. that was not a definite No.

I would soon meet her again at a party at an officer's house They were playing a rope trick game where two people were tied up. Each person has a three foot rope with each end tied from each wrist. The rope is looped inside the one tied to the other person's two wrists. The trick is you are supposed to get loose from the other person without untying from either wrist. You can imagine the contortions you could go thru trying to get loose. In fact you can now pull it up on YouTube and see it…. as the game is explained there. I never admitted it at the time but I knew the trick to getting loose which is actually simple and does not require either person to even touch the other. Why would I spoil the opportunity to get close to someone I was attracted to by quickly solving the puzzle. Besides there was another single Lieutenant there who had his eye on Mary Ellen and I could see he was getting extremely jealous of my close proximity. I pretended to be stumped at the solution until our host showed everyone there how easy it was to escape.

The next time I saw Mary Ellen, she was working at the Officer's Club. I believe she was taking up money as the cashier for the lunch buffet. She also worked in the Club office and that is where I started seeing her on a regular basis. I was still doing the monthly Club Calendar and it was printed in the office on their memo-graph machine. Mary Ellen and the other lady who worked in the office said they always had trouble running that machine. Earlier in this book I said that machine would become an important object in my life. Since I had no problem running it and I needed to print my calendars it became my reason to see Mary Ellen on a regular basis. I guess we became good friends long before we dated. After her divorce was final she dated other guys and I dated other girls. We still managed to see each other a lot. I got to know her son and spent time with them both. He was only one and a half years old.

I soon knew she was the girl I wanted to marry so I started planning when I wanted to propose. It had to be on a day which we would remember. I thought about Valentine's Day but everybody does that so I chose Saint Patrick's Day. She said she was Irish Catholic so that seemed appropriate. I worked on it for a week or maybe two. I bought a big fake ring and three Cracker Jacks boxes. I carefully steamed two of them open. In one I took out the cheap prize and replaced it with the large fake ring. In the other box the prize package was replaced with a note which read "WILL YOU MARRY ME"? And I included the engagement ring (which I had bought before January …. because I had also considered New Years Eve). The problem arose when I showed up at her house on the evening of March 17 to find she had another date. He was still at the house when I arrived so I decided to see if I could outlast him. If he had not left before midnight I guess my next proposal date would have moved to April Fool's day or maybe my birthday on April 4. The problem was solved because he left. We had Cracker Jack's. She found the fake ring but I don't think she thought it funny. Opening the second box and seeing the shape in the prize package she said something like "This must be my night for rings". Then she saw the note ….. and after a scary pause said yes.

WEDDING DATE IS SET

We decided to not have a long engagement and set our wedding date for June 24th. The Base Chapel was actually close to my office so we booked our wedding there. The rehearsal dinner and the reception were both set up at the Officer's Club. We met with the Catholic Priest, Father Bradel about performing the ceremony. I began calling him "Brutal Bradel" after he and I had a run-in over some papers he wanted me to sign. I refused because they were outdated Catholic stuff which had references to Masonic Organizations like they were part of some evil empire. I cannot remember exactly the wording but Masons were listed with some unfavorable names of things I would never even consider being a member of.

THE VATICAN BUT I CAN'T

Father Bradel even had the Military Ordinariate of the Catholic Church who at the time was Cardinal Francis Spellman call and talk to me. He was Archbishop of New York and not used to hearing an answer he did not approve of, especially from a half Jewish guy from Alabama. He said, "It's just a piece of paper which will probably be filed away in a drawer". I replied that it appeared to be an official document and even if it was just a formality … if it has something wrong it needed to be changed. I won't sign unless it is changed. We had a long talk and I finally told him that we were going to get married even if it were by a Methodist Minister, a Jewish Rabbi or a Justice of the Peace. Accept the fact I would not sign and let Father Bradel perform the service, especially if the Catholic Church wanted to try and keep Mary Ellen and her family happy with me marrying her. He finally said okay. I often wonder if changes were ever made. I rather doubt it.

Several friends did not try to talk me out of getting married but they did ask if I knew what I was doing by taking on …. not only the responsibility of being a newlywed but also starting out with a family. I said that I knew that I loved her and I loved her child and that soon she would be my wife and he would be my son. My whole life I have always been able to handle taking on responsibility. Even Father Bradel, in some one on one conversations, asked me if I knew how to start out marriage being a parent. I think I gave him a pretty good answer as we left the chapel after the ceremony. He was standing by the door as we exited …. I stuck a cigar in his pocket with the label "IT'S A BOY".

The chapel was not air conditioned. It was so hot that when we were on the kneel-er during the ceremony, my perspiration dripped down and I wrote "I LOVE YOU" on it.

After the reception at the Officer's Club, we left on our honeymoon for a week in the Pocono Mountains. An interesting side note to our honeymoon trip. I started saving quarters the day after we got engaged. In three months I saved enough quarters to pay for everything we did with quarters. Believe it or not I still have a few of those left and it has been 55 years. Before we drove back to Harrisburg we went to McGuire Air Force Base in New Jersey to view our military quarters which we would be moving to. I will explain how that happened in a future chapter.

OLMSTED AIR FORCE BASE CLOSING

Long before I had even met his older daughter (or even knew he had one) Colonel Torresson had given me and a Major Kramer from Group Headquarters a special assignment. The Colonel had heard thru the grapevine that Olmsted might be on the list of base closings. President Kennedy had announced the closing of 73 military establishments in 1961. That practice was continued by Defense Secretary McNamara and President Johnson. No one was sure of which bases would close but Colonel Torresson wanted us to look for a new home just in case. For each of our TDY orders to visit a base within our seven state area the reason for our visit was marked Classified. Now Major Kramer was a character. All of the secretaries were scared to death of him. He could be abrupt. If someone said “Good morning Major Kramer, how are you today” His reply would be, “Why do you ask? You don’t really care”. It was not just the secretaries …. many Officers, NCO’s and Airmen also found him hard to get along with because he often seemed in a foul mood. Lieutenant Glick and I were the exception. We got along fine with him because we would give it right back. He and I traveled well together. He asked all the right questions on the bases we visited and I handled details. We would visit a base, meet with many different people about facilities, buildings available, services offered and anything a tenant unit might need. We would give a report from each base to Colonel Torresson. We probably visited six bases and some more than once. We all concluded that McGuire Air Force Base in New Jersey would be our choice. They had two large buildings which were vacant (much larger than the ones we had at Olmsted). The second building had an open first floor which I could turn into a warehouse for my growing supply of displays and the parking lot was large enough to park my Hound Dog Missile and Space Capsule. Before the orders came for Olmsted to close we had already found a new home and made arrangements to move.

Viet Nam was beginning to heat up in 1966 so to me closing military facilities made little sense. My old housemate Charles Earnhart was in Nha Trang in late 1965 and we often talked on a Watts line at night. Through him I had sort of kept up with many of the things we never saw on the nightly news. Things were calm where he was but being with the Office Of Special Investigation he knew where the dangerous areas were. By the time I was thinking about getting married I had already lost some friends that I knew over there and I was pretty sure I did not want that as my next assignment. It was also around the same time that my position as Chief Of Advertising And Publicity was deemed critical and I lost my date of separation. My ID card now said “Indefinite”. As a fully qualified Staff Information Officer there were not that many of us, so I knew my name would creep up the list to be sent to Viet Nam. I called my friends in the Information Office at the Pentagon. I asked them if they could find out where I was on the list.

How accurate they were I will never know but when they said that my next permanent change of station could be Saigon it grabbed my attention. At the time Saigon was pretty tame and most likely I would have been in a safe office giving out casualty information to the press but with me considering getting married and knowing I would start out with a child I wanted to try my best to avoid this possibility.

An opportunity appears with an EXTRA DUTY. When Lieutenant Glick was on leave I filled in as Chief Of Personnel. I was going through regulations one night looking for information on a set of orders I was trying to cut. I ran across a chapter which said, “No Career Officer can be subjected to two non voluntary permanent changes of stations in a row without being given the opportunity to get out of service”. On the first one, you serve the full tour but you submit paperwork saying it is non voluntary. I am sure that was not the exact wording but you get the gist.

I sat there trying to comprehend what I had just read. I am thinking ….. they are closing Olmsted so when I move to McGuire that will be PCS (permanent change of station). If I say in writing that the move is non voluntary, I will still have to go but according to the regulation, I could not be given a second PCS move. I must have reread it four times. I finally figured out why the regulation had been written. It followed chapters referring to Viet Nam and Laos which were both remote assignments. It did not get much press but we were losing troops in Laos as well as Viet Nam. The regulation was written to keep anyone from being sent on two non voluntary remote assignments (where you could be killed) in a row. But the regulation did not say that. I think I stumbled on a loophole in the way a regulation had been written. I immediately typed up a letter referencing the regulation number, paragraph, etc. saying that I did not want to be moved from Olmsted AFB to McGuire, AFB. I would go but it was non voluntary.

I waited until Lieutenant Glick got back before I submitted my letter. I wanted it to come through him as the Chief Of Personnel instead of me as Acting Chief. He laughed and said "I think you may have found something here". The paperwork went up through channels. I wondered how my boss and soon to be father-in-law was going to react. He was not upset at all about me wanting to get out of service. He knew I did not want to make the Air Force my career and besides he was a Command Pilot. To him I was never really in the "real" Air Force. Also the joke I have used for many years and had used it with him was "When you marry the bosses daughter you expect to inherit the business and I found out he did not own the Air Force so I wanted out".

MOVING TO McGUIRE AIR FORCE BASE

Since we had a head start on finding a new base to relocate we were way ahead when the announcement came to close Olmsted. Since I had been with Major Kramer on all the sites we had visited I became the point person to set up all our contacts at McGuire. Lieutenant Glick and I started making TDY (temporary duty) trips to New Jersey to make sure everything was ready when we made the actual move. We worked with the Civil Engineers to get both buildings ready and Base Housing to set up quarters for everyone needing them. That was how I already had quarters assigned before Mary Ellen and I got married.

Even though Howard and I were working out of some temporary offices at McGuire we were still doing our jobs at Olmsted. I was still setting up advertising and publicity with all of our Detachments. Howard was handling all personnel matters. One which he had early on was the reassignment of a Lieutenant Colonel who was coming in new to Group Headquarters. Rather than have him report to Olmsted and then move within a couple of months Howard brought him straight to McGuire. We had to put him in Howard's temporary office and move Howard in with me. As you can imagine trying to keep up with our assignments at Olmsted as well as what all we were doing at McGuire we were quite busy. Neither of us hit it off well with the new Lieutenant Colonel. Even though he would be taking over the Educational Division when the Group finally got moved he thought he should be in charge (as the ranking Officer) of Howard and I at McGuire. He was a pain in the ass. We knew what we were doing and did not need him questioning everything we did especially since he knew nothing about what either of us had to accomplish.

Needless to say Howard and I found excuses to go back to Olmsted as much as we could. I had been working for months on a promotion with Millersville State College in Pennsylvania. They were having an Armed Forces Day where they wanted someone from each branch of service to come and speak. I had a bright idea to try and make the Air Force speaker the high point of the day and at least try to outdo the other branches. I asked Colonel Torresson if he would be willing to be our speaker (this was way before I was engaged to his daughter). My idea was to get the Helicopter Squadron at Olmsted to fly the Colonel in and land as close to the speakers platform as possible. He would hop out make his speech and fly off into the sunset. He liked the idea ...so I set about making it happen.

Millersville loved my proposal and agreed to mark off a large landing area next to the speakers stand. They would even get the local fire department to have a truck standing nearby. The Helicopter Squadron said I needed to go with them maybe the day before to do a practice run. What they did not tell me was they were going to make it also a training mission for several new pilots. When I went with them, we flew to Millersville made the landing, stayed as long as the speech was supposed

to last and took back off. The part they failed to tell me was that when we got back to Olmsted they would be practicing all kinds of take-offs and landings. I had the pleasure of at least 14. After awhile you just want to be back on the ground.
Anyway I am excited because the papers and TV are not only going to cover our event but they had actually shot film and photos of our practice run. We should get great publicity. Let's see the Navy, Marines and Army top us. The next day the Colonel called to say he is not feeling well and the new Lieutenant Colonel (who was now visiting TDY from McGuire to learn about his duties in the Education Division) had volunteered to take his place and give the speech. I called Howard and asked why he never told me our "friend" was in town. He said that he figured they would keep him busy learning his new job and we would not see him. Wrong ….I was supposed to go back with the chopper to the college. So at the last minute...I had something come up which I needed to stay and handle. Here is where the story gets good.

When the chopper is almost to the college it develops engine trouble and auto rotated to the ground. It landed in a plowed field next to the college and tipped over. From the spectators at the college (and TV crew and photographers watching) it appeared to crash. Smoke even came up. The pilot suffered a hand injury but no one else was hurt. I have to admit the Lieutenant Colonel took charge. He commandeered a guy on a tractor and had him drive right up to the speakers platform. We made the news ….. as I had told Colonel Torresson many times Publicity both good and bad is still getting our name out front. We had several area Recruiters attending the event and they made sure that the Lieutenant Colonel got back to Olmsted. He would always say that Lieutenant Glick and I planned it all to get rid of him. Wish we had been that smart.

Our two building were coming along nicely. Either Howard or I one supervised most of the reconstruction of offices. In the main building we had Colonel Torresson a large office. Howard did not have one which was closed in but it was in an alcove directly next to the Colonel's. For some reason the reconstruction crew could not make changes on the first floor because most all of the offices were already set up. Howard wanted to have one with a door. Since he and I both had that Dog Robber instinct the two of us decided to build him an office,

I had driven from Olmsted in a Ford Econoline Van because I had already started setting up my warehouse and I brought a load of displays. We took the van to the base dump. You would not believe what all the Air Force throws away. We found airplane parts, many of which I picked up to turn into displays. We found two by fours, and old door (still in a frame). There was Sheetrock (maybe five sheets) and even some wood paneling. We loaded up at least three trips worth of "junk". We went to a carpentry shop on base and talked them into letting us have nails, Sheetrock mud and tape, and even borrow a couple of hand tools and a circular saw. In one long night, Howard and I built him an office. When the Colonel saw it he was so impressed he wanted us to put a panel wall in his office like we did in Howard's. We had enough sheets to do his front wall the same as Howard's. I guess the icing on the cake came when the painters came in just before we were to move everyone in. They were looking at plans for the building and said, "This is strange. This office does not show up on our plans". His outside wall still got a coat of paint.

By around the middle of August of '66 we had everything set up at McGuire. My official date was July 16th but I had quarters there starting on June 26th (two days after I got married). Everyone seemed happy at our new location. We were close to New York and travel seemed easier from New Jersey. The base would occasionally provide a bus to take anyone to New York City and back to attend a Broadway Play. I think we took advantage of it at least once.

After we got to McGuire, Mary Ellen said I complained all the time about her dad giving me every dirty job. I did seem to get more than my share. I always felt that in his effort to not show favoritism toward his son-in law he went a bit overboard on giving me extra duties. Thinking back now after all these years he used to give me all those duties before I even knew he had an older daughter mainly because he knew I would get them done. After we were married the Colonel did occasionally bring up that letter from the Guidance Counselor. Even after I got out of service and he had retired it was something I heard about for several years. I never explained it then and don't plan on doing it now. I think his fantasy was far better than any explanation I could have come up with.

One of those extra duties was not so dirty but it was one which was complicated. The two star General in charge of all Air Force Recruiting was upset about our recruiters in the field having so many traffic accidents. The 3502 had more than the other six groups. Colonel Torresson said we need to do something about it. He called a large group of us in Headquarters and asked for suggestions …. then he told me I was in charge of coming up with a plan. He said that I could get ideas from anyone on his staff but put something on paper which he could show the General. Since my fishing buddy Sergeant Hodge was in charge of the Motor Pool I knew he had all the facts about accidents in our group so I asked him to help me. I threw out an idea which he liked.

GOVERNOR'S VEHICLE SAFETY AWARD

I had so much success in Amarillo when I came up with the trophy for the Student Squadron Beautification Award that I thought how can a trophy be used for Vehicle Safety. Then it hit me. We have Recruiters driving cars in Seven States. Why not not have each state compete and have the Detachment with the best overall driving record receive a trophy from the Governor of his state. I bet I could get some good publicity mileage out of the Governor awarding the trophy to the Detachment Commander and maybe he could include one of his recruiters who had the best record. Thinking of it is one thing, selling it is another.

First I needed a trophy. I wanted a big one so I could send a photo of it to each Governor trying to get them to commit to participating in the program. I knew getting seven different politicians to agree on anything would not be easy. I put out the word to my Ad and Pub contacts in each of our Detachments. All of them were super resourceful. Sure enough one of them had a COI (center of influence) who owned a trophy shop. He loved our idea and said if we gave him credit if the winner was from his state he would donate a trophy. I believe he was in Virginia. He came up with a trophy about four foot high. It had a plaque where we could add names of winners. He even made the suggestion we award it every six months instead of a year so we would get more mileage out of the program. As soon as it arrived I immediately took photos and started writing Governors. I wrote what I thought a good letter. I played on the ego of the Governor pointing out the publicity I thought we could generate and how promoting safe driving in each state always looked good with the public. Answers came back fast from everyone except Maryland. I wrote him again. This time I got an answer probably from someone on his staff saying that his schedule would not allow him to participate in programs like ours because then everyone would be asking for him to support this and that promotion.

I wrote another letter marked Personal And Confidential. It was in a large manila envelope with a note on the outside saying Contains Personal Information From Six Governors. I told him it was sure going to look funny when the first winner was announced and we showed in our Press Release which of our states participated and which ONE did not. When he got the letter he called personally and said that he never saw the first correspondence and he definitely thought our program was a good one and we could count on him. Don't you just love politicians.

The program was successful. The 3502nd went from last to first in vehicle safety. The General was pleased which made everyone down the chain of command also pleased. Our first winner was from Pennsylvania so I got to make a trip back to Harrisburg where Governor Scranton posed for photos.

A funny story goes along with that photo shoot. I used to love seeing Press Photographers with their Graflex Speed Graphic Camera and big Flash attachment. I wanted one for my Division. In checking I found that we were not authorized to have one. I put out the word and one of my Dog Robber contacts came back and said he found one which was surplus at the Army Depot. I went and got it on a Hand Receipt much the same way I acquired many other things. Now we needed to learn to use it.

The base Photography Lab said they would teach a class but needed at least three students before they could set one up. That meant two of my NCO's and I. The problem was I traveled so much I missed over half our classes. The day of our final we were supposed to take a portrait shot of the instructor. I got everything set up, the view finder, the focus, the plate in correctly now just press the shutter. Imagine my embarrassment and the laughing of all who were present when the flash

bulb flew across the room. Luckily one of my NCO's shot the photo of Governor Scranton. I was in the photo and not behind the camera.
I eventually did improve and with our Press Camera my staff and I at least looked like we knew what we were doing. The photo on the cover of this book was shot with that camera.

A DAD AGAIN

I barely got used to being a dad when Mary Ellen found out she was pregnant. The medical facilities were not exactly First Class. In fact everyone at McGuire used Walston Army Hospital at Fort Dix which was next door to McGuire. The Maternity Ward was an almost "Do It Yourself" program. I want to rephrase that. It was not "Almost". It WAS a Do It Yourself Ward. If we had a DIY TV Network back then, they could have done a series. The ward was filled with pregnant women who's husbands (or boy friends) were deployed in Viet Nam. When I went to visit Mary Ellen and slipped in food I was a pretty popular guy. Several times I sneaked in, found a closet and put on a doctor's robe and made my rounds. One night a young girl in labor, who was on a gurney in the hall asked me to see how far along she was. I just said, "Few more centimeters to go and kept walking". Mary Ellen gave birth to a beautiful baby girl. They had left my wife alone and never gave her anything so she was on her own when the time came. As I said it was a Do It Yourself ward. The hospital bill was $5.25. My wife thinks she was overcharged. The patients were even expected to get up and go get their own food trays. With that and the fact the meals did not look edible is why I tried to slip in food. Our new edition we named Sandra Elaine Capps. Mark now had a little sister.

BIG BROTHER

Mark took over as "Big Brother" immediately. He did not act jealous and liked holding the new baby. He did like to get away from the crying every now and then. One thing he enjoyed was going with me over to my warehouse. In our new building the whole first floor was open and it became a treasure chest of things a small boy found of interest. We had everything from parachutes to flight helmets to small model planes. One Saturday morning he was with me while I checked on a display which was to be shipped out on Monday. I noticed him playing with a large trash can. It was lying on it's side. He would lay across it on his stomach and rock back and forth. The idea hit me that maybe I could make it into some kind of toy which kids could play with. The can was made of extra thick cardboard material which did not bend. At first I thought maybe some kind of seat on top with wings on the sides which would keep it from rocking all the way over. Sort of a fat airplane. I could easily make a propeller for the front. Then it hit me… Mark was having fun just rocking back and forth. While he was playing I made a drawing on a piece of paper.

The "ROCK-A-CHOO-CHOO" Was Born

All I had to do was build it. For the next two weeks in my spare time I built Mark a toy for Santa to bring. Christmas was almost there before it was finished. It turned out much bigger than I intended and it would not fit through the door to our quarters. Mary Ellen's brother Skip and I had to take the door off on Christmas Eve (in the snow) for Santa to have it appear on Christmas morning.

It was built so well it survived a move from New Jersey to Alabama (and three moves after that). It is too hard to describe but years after it was built it appeared with a set of plans in Southern Living Magazine. It was multi-functional as the front part (which was the trash can) also served as a toy chest. It was named the "MARK 4" and became the hit of the neighborhood with the kids. Mark shared it with his little sister. When she got old enough he taught her how to get into it and rock back and forth.

Made from scraps found at the dump at McGuire AFB

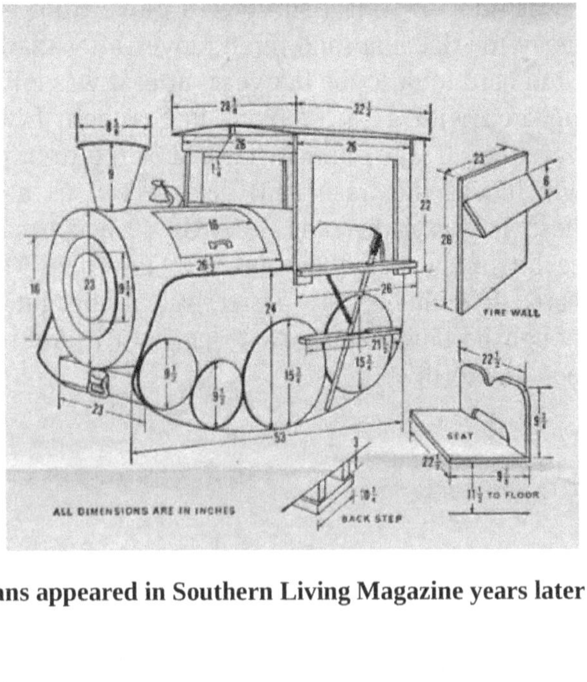

Plans appeared in Southern Living Magazine years later

AS VIET NAM HEATS UP McGUIRE GETS BUSY

The new Military Airlift Command (MAC) got extremely active beginning in 1966 and by 1967 McGuire was home to up to a quarter of all C-141 Starlifter aircraft in the world. It had replaced the slower propeller driven cargo planes and was heavily used throughout the Viet Nam war. It was not unusual to see return flights arriving with flag draped caskets of the casualties from the war. While most were sent to Dover Air Force Base in Delaware many landed at McGuire. It was a moving experience to watch.

Around the first of September 1966 I received notification of promotion to Captain. Because of being extended and losing my date of separation I had acquired enough time for promotion. This would certainly help my base pay both during the remainder of my time at McGuire but also help when I became discharged. We were allowed 30 days leave each year and we could carry over up to 60 days. I was planning on having the full 60 days after I got out so I would have income and at least two months to find a job.

Did I ever consider the Air Force as a career? The answer is yes. I loved my job. If I could have kept working as Chief of Advertising and Publicity and stayed in Recruiting, I would have stayed in the Air Force until retirement. I admired Colonel Dean Hess, my counterpart at Recruiting Service Headquarters. If I could have seen an opportunity of moving up through the ranks to one day being promoted to the position he occupied that would have appealed to me. The Air Force did not operate that way. I was a fully qualified Staff Level Information Officer. There were not that many of those, so I could have been assigned to any number of jobs ….. several of which at that time happened to be in Viet Nam.

It was a long time coming. We had been at McGuire almost a year. I was now a Captain. Colonel Torresson got a call from a Colonel at Personnel Headquarters in Denver. He said,
" There is a Captain Capps in your Headquarters Group. Does he know an Air Force Regulation is having to be changed because of him"? Colonel Torresson never told him that I was his son-in-law but he said "I think he is aware of that". I was relieved from Active Duty on 30 June 1967 but remained on Inactive Reserve status until I was discharged as a Captain in May 1972. I was contacted several times to get me to join an Active Reserve Unit but declined the offers. I probably could have done that as a Major because when I got out I was told that if recalled to active duty because of Viet Nam I would come back with that rank.

Well there you have it ….. a little over five years of experience which could never be gained through another source. I would encourage anyone to join the Air Force. If you are willing, they will allow you to take on any responsibility you can handle. Before Mary Ellen's dad, Colonel Thomas S. Torresson, Jr. passed away, he told me many times that I should try to put it all on paper some day. He said that I had more weird stuff happen in five years than he had when he retired with thirty.

Be it right place at the right time or wrong place at the wrong time …. you be the judge.

Around the time I was teaching art classes in Amarillo I drew up a cartoon of a Second Lieutenant. He was holding a long list of Extra Duties. I put it on the front of a greeting card with the caption "Everything I Am Today I Owe To You". On the inside it said "I'll Get You For This"! I sent it to my Air Force Recruiter. When I was home on leave I stopped by his office to say hello.

There, framed on his wall was my card.

ABOUT THE AUTHOR

William Peary Capps was born in Birmingham. Alabama. He graduated from Auburn University in 1962 with a BA-A degree in Advertising Design.

He served in the U.S. Air Force for five years active duty and five years inactive, discharged in 1972 as a Captain. He held the position of Chief of Advertising And Publicity for one of the seven Recruiting Groups.

He had a 27 year career with Southern Progress Corporation (formerly Progressive Farmer Company). Bill held several positions over the years but served the longest as Vice President and Circulation Director for all the publications, which included magazines such as Progressive Farmer, Southern Living, Creative Ideas, Southern Accents and Cooking Light. During that period his department sold 98 % of all subscriptions by direct mail. In 1995 he retired from Time, Inc. which bought the corporation in 1985.

He served six years on the Board of Directors for the New York based Direct Marketing Association. Bill was a guest speaker at various functions all over the U.S.

Upon retirement he concentrated on his one man company, Cappsize Creations, consulting with many different publications and various other companies and organizations. Cappsize also did artwork and design and owned several rental properties which included residential, vacation and currently business. When asked if Cappsize Creations had been successful, Bill replied "Well I have never had an employee complain and it has paid taxes for 51 years".

Polaroid shot from 100 Ft. ladder by Lt. Capps after dare from the Ramat Gan Fire Chief

Actual Hound Dog Missile In Flight

Hound Dog Missile and Space Capsule On Display At A County Fair

**Bill & Mary Ellen Capps
on their wedding day
June 24, 1966**

Made in the USA
Monee, IL
04 November 2026

67386405R00213